Advanced Methods for Sustainable Farming: Principles and Practices

Advanced Methods for Sustainable Farming: Principles and Practices

Shashi Shekhar Prasad

RANDOM

RANDOM PUBLICATIONS
NEW DELHI (INDIA)

Advanced Methods for Sustainable Farming: Principles and Practices

ISBN 978-93-5111-587-8

Published in 2015 in India by

RANDOM PUBLICATIONS

4376-A/4B, Gali Murari Lal, Ansari Road
New Delhi-110 002
Phone : +9111-43580356, 011-23289044, 011-43142548
e-mail: sales@randompublications.com,
info@randompublications.com, randomexports@gmail.com

Reprint 2021

Type Setting by : Friends Media, Delhi-110089
Digitally Printed at: Replika Press Pvt. Ltd.

Preface

Agriculture is faced worldwide with an accelerating transformation. Advances in science and technology, shifting consumption patterns, continuing population growth, trade globalization, frictions in subsidy regimes, and the impacts of local and global environmental change converge and lead to new and serious risks to agricultural production systems and producers. In light of these changes, sustainability of the agri-food system and farming as a multi-functional enterprise is of increasing importance. Successful short- and long-term adaptation to endogenous and exogenous biotic and abiotic, social and economic forces requires access to information and indicators on the current situation and possible future trends in the ecological, social and economic domain of sustainability.

Although sustainability measurement and assessment has been on the research agenda of many organizations for quite some time, their actual impact on policies, practices and outcomes on the ground beyond pilot initiatives has been limited. Our hypothesis and strong belief is that if placed in the right context, and their value-added convincingly demonstrated, sustainability measurement and assessment has the potential to reduce long-term risk and advance the sustainability agenda well beyond the current state.

I would like to thank my team for standing beside me throughout my career and writing this book. My special thanks go to "Random Publications" who have published the book.

– Shashi Shekhar Prasad

Contents

1

Agriculture: Farmers and Producers

The issues that are important to farmers across the country include stewardship, profitability, sustainability and the health of their community. And those are the topics to be covered in Kansas in an upcoming agriculture roundup.

The practice of sustainable agriculture is built upon soil fertility and the protection of soil health. For centuries, farmers around the world have been employing these basic techniques to keep their soil productive. Early in colonial American history, George Washington was using cover crops and manure applications on his farm. American farmers, who did not carefully tend their soil, eventually "wore it out."

The Westward Expansion partly reflected the lowered productivity of eastern lands and the search for new farmland farther west. The first publications from the Kansas State Agricultural College appeared in the late 1880's, at a time when Kansas soils had been tilled for less than fifty years. Soil scientists were well aware at that time that the long term sustainability of farming depended upon the use of legumes for fertility and the addition of organic matter for tilth. As Kansas farmers began to report declining fertility levels in the early 1900's, scientists cautioned that farmers must not rely solely upon the natural fertility of the prairie soils.

Between 1900 and 1950, Kansas State researchers guided farmers in the use of legumes and the preservation of soil organic matter. With the advent of inexpensive nitrogen fertilizers after World War II, traditional soil management techniques took a back seat to the use of commercial products that were easy to use and often provided greater economic returns in the short-term. At the beginning of the twenty-first century, farmers are becoming increasingly conscious of the importance of soil health, water quality, and energy conservation.

The rising cost of nitrogen fertilizers has revived interest in nitrogen-fixing legumes. Excess phosphorus levels in surface water indicate a need to emphasize soil conservation and the careful use of manure resources. Traditional soil management practices continue to be vital for the sustainable agriculturist and are regaining an audience with conventional agriculture. This review of Kansas State University soil publications profiles the research and recommendations of Kansas scientists during the early to mid-twentieth century. Generally, recommended practices such as crop rotations, manure use, cover crops, and other sources of fertility are considered including the shift to commercial fertilizers in the 1950's. The special consideration of western Kansas soils is treated separately. Limited rainfall in western Kansas affects the research conducted in that part of the state and alters the use of traditional soil health practices.

STRUCTURE AND FERTILITY OF SOIL HEALTH

Within every aspect of Kansas agriculture, healthy soil is a key element. Its structure and fertility provide the basis for all crop and livestock production. Throughout the historical publications of Kansas State University from the late 1800's to the mid-1900's, researchers and educators have been concerned with the protection of this vital resource. In their 1918 publication, Soil Fertility, L.E. Call and R. I Throckmorton caution the Kansas farmer. "The soil is the most important source of wealth in an agricultural state. If it is maintained in a high state of productivity, by wise systems of soil management, the people prosper.

If its fertility is wasted through careless methods of farming, both the farmer and the state suffer". Forty years later, in 1956, Orville Bidwell echoes this same message with its promise of a precarious wealth. "Unlike most other resources, soil is inexhaustible if properly managed." Bidwell recounts the variety of Kansas soils each with a different waterholding capacity, permeability, response to fertilizers, and susceptibility to erosion.

DECLINE OF KANSAS SOILS

Throughout many of the earliest publications, the authors are clearly concerned about the declining condition of Kansas soils. By the early 1900's, much of the rich prairie soil in eastern Kansas had been farmed for 50 years. A 1903 publication from the veterinary department of the Kansas State Experiment Station states, "The fertility of the soil of the Middle states and the West is being rapidly diminished and if means are not taken to prevent it, the time is not far distant when it will be as necessary to apply artificial fertilizers to the soil as it is now in the East".

A few years later, chemists at the agricultural experiment station raised the same concern. "In the early history of Kansas no attention was paid to the composition of its soils except to boast of their inexhaustible fertility. The voice of the chemist has been lifted constantly, warning the people that this idea of possession of a fertility that is practically limitless is a delusion that can lead only to squandering of our natural resources, and to leaving posterity handicapped in the struggle for existence. Today he is seeing his warnings justified. People in many localities of the eastern part of the state are making enquiry concerning chemical analysis of their soils with reference to learning what fertilizers should be applied and to what crops their soils are best adapted".

By 1918, L.E. Call and R. I. Throckmorton attempted to put dollar figures to the losses in soil fertility. They estimated that the plant food removed from Kansas soils by wheat crops over the previous fifty-five years equalled a value of more than seven hundred million dollars. Even the wheat straw, which was regularly burned or wasted, had a value in plant nutrients of more than twelve million dollars.

The majority of the wheat products were both milled and eaten outside the state which Call and Throckmorton equated with the export of soil fertility. Call and Throckmorton credited the declining productivity of Kansas soils to five factors: depletion of soil organic matter, failure to grow enough acres of leguminous crops for nitrogen fixation, depletion of mineral nutrients, the lack of proper crop rotations, and the erosion of fertile topsoil. These five factors are the subject of nearly every soils publication prior to the advent of inexpensive nitrogen fertilizers in the mid-1900's. They continue to be the basis of soil health for every farming system regardless of the use of commercial fertilizers.

SOIL ORGANIC MATTER

The early Kansas State publications recognized the importance of soil organic matter to soil health. Replenishment of soil organic matter is the most basic step in addressing a number of other issues. According to Call and Throckmorton, organic matter holds the soil's store of nitrogen, provides good tilth, holds moisture, and provides food for the bacteria that make nutrients available to plants.

Soil studies in western Kansas spanning over 30 years of crop production give mention to the importance of organic matter. Progress reports in 1943 and 1957 indicate that as the organic carbon content of the soil has decreased, more power is needed for tillage, water intake is decreased, seedling emergence and root growth are hindered. The 1943 report also mentions that the presence of coarse, fibrous organic matter helped reduce wind erosion. Numerous publications cite frequent plowing and intensive cultivation as culprits in the loss of organic matter. Row crops depleted the soil more quickly than small grain crops. Without a regular practice of restoring organic matter, the levels of carbon in the soil would drop, threatening fertility and soil structure.

The 1918 publication, Soil Fertility, promotes barnyard manure applied to the soil as a primary source of organic matter. However, when sufficient manure is not available, the farmer should grow a crop to plow under to supply organic matter. These green manure crops may be legumes such as alfalfa, cowpeas, soybeans, clover, or sweet clover. Legumes would capture atmospheric nitrogen and fix it in the soil in addition to supplying organic matter. Non-legume crops such as rye, buckwheat, sorghum, and turnips were also cited as possible sources for organic matter.

Although most of the methods for increasing organic matter focus on the addition of green manure crops or barnyard manure, the conservation of all organic matter is addressed. Throckmorton and Call state that most wheat straw is either burned or destroyed after threshing. They recommend its use as feed and bedding for livestock.

Eventually the soiled bedding and any manure would be returned to the soil. They also suggest use of the straw as a surface mulch on wheat during the winter at a rate of 1-1.5 tons per acre. In 1962, farmers were again cautioned not to waste the organic matter provided by straw or stubble. In an effort to control weeds and increase yields, burning of wheat stubble had become a common practice. A series of studies in western Kansas produced data showing that stubble burning did not increase yields for subsequent crops. Not only did it present increased potential for wind erosion, it also decreased the soil's ability to absorb water.

NITROGEN AND ORGANIC MATTER OF SOIL

The supply of nitrogen in the soil was closely linked to the organic matter content. The practice of returning organic matter to the soil in the form of leguminous plants or animal manures was also the method for supplying nitrogen prior to the use of commercial nitrogen fertilizer. Legume crops, grown in rotation with other crops, could be worked back into the soil as a green manure crops or the legume hay crop was fed to livestock and their manure was returned to the soil.

Off-farm sources of nitrogen in the early 1900's included waste materials from the packing houses and inorganic compounds such as saltpeter, which was mined in Chili, and manufactured compounds of ammonium sulphate or calcium cyanamide. Researchers noted that "the purchase of nitrogenous fertilizers should be limited to the meeting of special requirements of certain conditions or crops...Nitrogen...is the (nutrient) most cheaply restored, since by the cooperation of clovers, alfalfa, peas, beans, and other legumes with bacteria that grow upon their roots the abundant nitrogen of the air in the pores of the soil is brought into organic combination. This means of adding nitrogen to a soil must never be lost to view...".

Through the first half of the twentieth century, researchers and farmers sought to understand the best methods for capturing nitrogen with legumes especially when those crops were being used for other purposes on the farm. In 1918, researchers were concerned that the nitrogen found within the crop roots and stubble might not be significant compared with the amount removed in a hay crop. They hypothesized that a certain amount of leaf loss during the haying process might be returning some nitrogen to the soil. Still they cautioned that the best practice was to feed the hay on the farm and return all manure to the soil. The 1939 publication on fertility studies at the Manhattan experiment station beginning in 1910 gave evidence that soil nitrogen could be increased even when the hay was removed. The average increase in the 5-9 year old hay plots was twice that for the 1-4 year plots. The researchers also found that the residual effects of the "nitrogen accumulating capacity" remained for eight to nine years following the breaking of the alfalfa sod that had been in alfalfa for more than two years.

But nitrogen was not the only concern when hay was taken off the farm. "If alfalfa is sold off the land it is one of the most soil-exhausting crops raised, while if it is fed on the farm, and the manure produced applied to the land, it is a conserver of fertility. The same argument applies to clover". Potassium, phosphorus, calcium and other mineral elements are taken up by the plants and must be cycled back to the soil as green manure or barnyard manure. When hay is sold off the farm, the phosphorus and potassium are more rapidly depleted than with a continuous grain crop.

Regardless of the use of green manure crops, any farm with livestock had a source of organic matter and nutrients in animal manure if it was used wisely. "If the livestock farmer properly saves and utilizes his manure he can maintain his soil in a high state of productivity, but the livestockman who feeds his cattle in woodlots along the banks of streams, and so wastes his manure, usually depletes the fertility of his soil more rapidly than the man producing grain only.

It was important to manage manure so that nutrients were not lost before they were cycled back to the soil. The seepage of liquid waste or urine, leaching of manure by rain and run-off water, and the "decay" of manure solids and the subsequent loss of nitrogen are all a result of poor handling techniques. Throckmorton and Call recommend that the most practical method of manure handling would be to feed the stock on the cultivated fields so that the manure is scattered by the animals and the nutrients are retained by the soil. In any case, the manure should be returned to the soil as soon as possible and long term, open storage of six months or more should be avoided. Manure could also be used sparingly as a top dressing on corn, kafir, or winter wheat where it would act as mulch to retain moisture. If there was not a large supply of manure, it was deemed better to apply the available supply lightly to more acres rather than a heavy application on just a few acres.

As a percentage of volume, the nutrients present in manure are small. "The figures for the fertilizing constituents are always low, but they are present in readily available form and the accompanying organic matter has itself a highly beneficial effect on the land. Even with these low percentages the total amount of plant food in the manure produced on a farm reaches very significant quantities".

PHOSPHORUS

Fertilizers were commonly used to provide phosphorus and other minerals long before nitrogen fertilizers were in general use. By 1918, the soils in eastern and southeast Kansas had such low stores of phosphorus that it was profitable to purchase phosphorus fertilizers. Grain farms lost phosphorus the most rapidly when it was exported off the farm with the grain. In order to retain the mineral, the farmer had to feed the grain to livestock and apply the manure to the croplands or else import a supply of phosphorus.

Alfalfa and other deep-rooted crops could be used to collect phosphorus from the subsoil. The nutrients then needed to be cycled back into the upper levels of the soil as either a green manure crop or manure from animals fed on the hay.. Commercial sources of phosphorus included bone, basic slag, rock phosphate, and apatite.

Bones were a valuable product rich in phosphorus and nitrogen. They could be ground raw but they were more commonly steamed before grinding. This processing made grinding easier, concentrated the phosphorus slightly and increased the availability to plants.

Basic slag was a by-product from a particular method of iron refinement. Minerals removed from certain iron ore contained high amounts of phosphate, which could be ground for use as a fertilizer. This was not a common product in America, being chiefly available in Europe. Primary sources of rock phosphate were found in Florida, South Carolina, and Tennessee during the early part of the twentieth century. Although it was most often used to produce superphosphate, it could be finely ground and used raw. Superphosphate was produced by treating the raw phosphate with sulphuric acid and was in use to some extent throughout the twentieth century.

Although superphosphate was more readily available to the plant, Kansas State researchers advised that the raw phosphate was usually a wiser choice. "In the application of phosphate fertilizers the farmer naturally expects and desires immediate results, which are secured by the use of superphosphate, but at the same time if larger quantities of phosphates can be applied in other less soluble and available forms at the same expenditure, the ultimate value of the investment may be much greater, as the phosphorus will remain in the soil and be rendered available by slow natural processes.

There was also indicated a positive relationship between the bacteria found in organic matter and the availability of phosphorus from raw phosphate.

The cheapest source of phosphorus is ground rock phosphate, and in this form it will be available for the use of crops if the soil is well supplied with organic matter from farm manures and legumes. Apatite, a crystal found in granite, was abundantly available in Canada. It needed to be converted to superphosphate to be used as a fertilizer.

POTASSIUM

Potassium occurs in the mineral or rock portion of the soil. Plant roots are able to access potassium directly from tiny soil particles, chiefly silt and clay. Potassium is also present in decaying crop residues and organic matter, indicating the need, once again, for returning plant materials to the soil. A traditional source of additional potassium was hardwood ashes. Muriate of potash, a processed form of potassium in wide use today, was also available early in the twentieth century but due to the chlorine content of the compound, its use was restricted with certain crops.

CALCIUM (LIME)

Calcium, while an essential element for plant growth, was usually not deficient in Kansas soils. However, its application in the form of lime, was a common soil amendment. "(Calcium) is generally present in all cultivated soils in sufficient quantity to supply fully the need of the plant. Yet even where this is the case, the soil may be greatly in need of liming. Lime is used, therefore, as a soil amendment not so much for its effect directly on the plant as for its effect on the soil, which indirectly affects the plant. Soils that are low in lime are said to be sour". The acidity of these soils could be neutralized by the application of calcium compounds. A 1918 publication recommended two tons of ground limestone per acre followed by one or two tons every five to six years thereafter.

The limestone should be worked into the bare ground six months to a year prior to seeding alfalfa or clovers. The effects of the limestone are slow and gradual but long lasting. The bacteria that live on the roots of alfalfa, sweet clover, and clover thrive in more alkaline soils so these crops respond well to liming. Crops considered less sensitive to acid soils were corn, wheat, timothy, and oats.

The change in soil pH improved the availability of phosphorus and potassium as well as improving the texture of the soil. "It is a well-known fact that soils well stocked with a supply of lime will be more productive under the same conditions of plant-food content than soils not so stocked...While soils use very small amounts of lime as compared with phosphorus and potassium, yet the presence of a relatively large supply of lime insures crop production. In the words of Hilgard, 'A lime country is a rich country'".

SOIL BACTERIA

Although the exact relationship did not seem to be clear, researchers in the early part of the twentieth century were writing about a correlation between humus or organic matter, bacteria, and fertility. A series of studies beginning in the 1890's showed crop yields in "direct proportion" to the bacterial content of the fields. In this same publication, the authors speculate that fertility depends to a large extent on bacterial activity and that by manipulating the bacteria of the soil, one might avoid the need for artificial fertilization. They proposed additional experiments to find ways to increase soil bacteria. A few years later, Walter King and Charles Doryland report on their studies of soil bacteria with this same basic assumption.

Assuming that bacterial activity indicated increased soil fertility, they set about collecting samples from various soil depths on plots representing a variety of tillage practices. They admitted tremendous variability in their data, which was collected over a period of only three months from March to June 1908. Nevertheless, they concluded that deep plowing, conveniently the tillage practice most commonly used at that time, increased bacteria levels and bacterial activity and decreased denitrification. They noted that bacterial activity increased with the temperature of the soil and decreased when the soil became saturated with moisture. Different species were predominate at different times and activity seemed to rise and fall with a regularity independent of moisture and temperature.

TECHNICAL PROCESSING OF AGRICULTURAL PRODUCE

Above we have already had occasion to note that writers on agriculture, in classifying systems of farming according to the principal market product, assign the industrial or technical system of farming to a special category. The essence of this system is that the agricultural product, before going into consumption (personal or productive), undergoes technical processing. The establishments which effect this processing either constitute part of the very farms on which the raw material is produced or belong to special industrialists who buy up the raw material from the peasant farmers. From the standpoint of political economy the difference between these two types is unimportant. The growth of agricultural technical trades is extremely important as regards the development of capitalism. Firstly, this growth represents one of the forms of the development of commercial farming, and is, moreover, the form that shows most vividly the conversion of agriculture into a branch of industry of capitalist society.

Secondly, the development of the technical processing of agricultural produce is usually connected intimately with technical progress in agriculture: on the one hand, the very production of the raw material for processing often necessitates agricultural improvement (the planting of root-crops, for example); on the other hand, the waste products of the processing are frequently utilised in agriculture, thus increasing its effectiveness and restoring, at least in some measure, the equilibrium, the interdependence, between agriculture and industry, the disturbance of which constitutes one of the most profound contradictions of capitalism. We must accordingly now describe the development of technical agricultural trades in post-Reform Russia.

DISTILLING

Here we regard distilling only from the point of view of agriculture. Accordingly, there is no need for us to dwell on the rapid concentration of distilling in large plants (partly due to excise requirements), on the rapid progress of factory technique, with the consequent cheapening of production, and the increase in excise duties which has outstripped this cheapening of production and because of its excessive amount has retarded the growth of consumption and production.

Thus, over 9/10 of the distilleries (accounting for over 4/5 of the total output) are directly connected with agriculture. Being large capitalist enterprises, these establishments lend the same character to all the landlord farms on which they are set up (the distilleries belong almost without exception to landlords, mainly to members of the nobility). The type of commercial farming under review is particularly developed in the central black-earth gubernias, in which are concentrated over 1/10 of the total number of distilleries in the Russian Empire (239 in 1896-97, of which 225 were agricultural and mixed), producing over a quarter of the total output of spirits (7,785,000 vedros in 1896-97, of which 6,828,000 at agricultural and at mixed distilleries).

Thus in the area where labour-service predominates, the commercial character of agriculture most frequently (as compared with other areas) manifests itself in the distilling of vodka from grain and potatoes. Distilling from potatoes has undergone a particularly rapid development since the Reform, as may be seen from the following data relating to the whole of the Russian Empire:

Thus, with a general twofold increase in the quantity of crops distilled, the quantity of potatoes used increased about 15-fold. This fact strikingly corroborates the proposition established above that the enormous increase in the potato area and crop signifies the growth of precisely commercial and capitalist farming, along with improvement of agricultural technique, with the replacement of the three-field system by multi-field crop rotation, etc. The area of the biggest development of distilling is also distinguished for the biggest (in the Russian gubernias, *i.e.*, not counting the Baltic and the western gubernias) net per-capita harvest of potatoes.

Thus in the northern black-earth gubernias the figures for 1864-1866, 1870-1879 and 1883-1887 were 0.44, 0.62 and 0.60 chetverts respectively, whereas for the whole of European Russia (50 gubernias) the corresponding figures were 0.27, 0.43 and 0.44 chetverts. As far back as the beginning of the 80s the *Historico-Statistical Survey* noted that "the region marked by the greatest expansion of potato cultivation covers all the gubernias of the central and northern parts of the black-earth belt, the Volga and Transvolga gubernias and the central non-black-earth gubernias" (*loc. cit.*, p. 44). The expansion of potato cultivation by landlords and well-to-do peasants means an increase in the demand for hired labour; the cultivation of a dessiatine of potatoes absorbs much more labour than the cultivation of a dessiatine of cereals and the use of machinery in, for example, the central black-earth area is still very slight. Thus, while the number of workers engaged in the distilling industry proper has decreased, the elimination of labour-service by the capitalist system of farming, with the cultivation of root-crops, has increased the demand for rural day labourers.

BEET-SUGAR PRODUCTION

The processing of sugar-beet is even more highly concentrated in big capitalist enterprises than distilling is, and is likewise an adjunct of the landlords' (mainly noblemen's) estates. The principal area of this industry is the south-western gubernias, and then the southern black-earth and central black-earth gubernias. The area under sugar-beet amounted in the 60s to about 100,000 dess., in the 70s to about 160,000 dess.; in 1886-1895 to 239,000 dess., in 1896-1898 to 369,000 dess., in 1900 to 478,778 dess., in 1901 to 528,076 dess. (*Torgovo-Promyshlennaya Gazeta*, 1901, No. 123), in 1905-06 to 483,272 dess. (*Vestnik Finansov*, 1906, No. 12).

Hence, in the period following the Reform the area cultivated has increased more than 5-fold. Incomparably more rapid has been the growth of the amount of sugar-beet harvested and processed: on an average the weight of sugar-beet processed in the Empire in the years 1860-1864 was 4.1 million berkovets; in 1870-1874—9.3 million; in 1875-1879—12.8 million; in 1890-1894—29.3 million; and in 1895-96 and 1897-98—35 million. The amount of processed sugar-beet has grown since the 60s more than 8-fold. Hence, there has been an enormous increase in the beet yield, *i.e.*, in labour productivity, on the big estates organised on capitalist lines.

The introduction of a root-plant like beet into the rotation is indissolubly linked with the transition to a more advanced system of cultivation, with improved tillage and cattle feed, etc. "The tillage of the soil for beetroot," we read in the *Historico-Statistical Survey*, "which, generally speaking, is rather complicated and difficult, has been brought to a high degree of perfection on many beet farms, especially in the south-western and Vistula gubernias. In different localities, various more or less improved implements and ploughs are used for tilling; in some cases even steam ploughing has been introduced".

This progress of large-scale capitalist farming gives rise to quite a considerable increase in the demand for agricultural wage-workers— regular and particularly day labourers—the employment of female and child labour being particularly extensive. Among the peasants of the neighbouring gubernias a special type of migration has arisen, known as migration "to sugar". It is estimated that the complete cultivation of a morg (= 2/3 dess.) of beet land requires 40 working days.

The *Combined Material on the Position of the Rural Population* (published by Committee of Ministers) estimates that the cultivation of one dessiatine of beet land, when done by machine, requires 12, and when by hand 25, working days of males, not counting women and juveniles. Thus, the cultivation of the total beet area in Russia probably engages not less than 300,000 agricultural day labourers, men and women.

But the increase in the number of dessiatines under beet is not enough to give a complete idea of the demand for hired labour, since some jobs are paid for at so much per berkovets. Here, for example, is what we read in *Reports and Investigations of Handicraft Industry in Russia*.

"The female population both of the town, and of the uyezd" "think highly of work on the beet fields; in the autumn the cleaning of beets is paid at 10 kopeks per berkovets, and two women clean from six to ten berkovets a day, but some contract to work during the growing season as well, weeding and hoeing; in that case, for the full job, including digging and cleaning, they get 25 kopeks per berkovets of cleaned beets." The conditions of the workers on the beet plantations are extremely bad. For instance, the *Vrachebnaya Khronika Kharkovskoi Gubernii* cites "a number of exceedingly deplorable facts about the conditions *of those working on the red-beet plantations*. Thus, the Zemstvo physician, Dr. Podolsky, of the village of Kotelva, Akhtyrka Uyezd, writes: 'In the autumn *typhus* usually breaks out among *young people employed on the red-beet plantations of the well-to-do peasants*.

The sheds assigned for the workers' leisure and sleeping quarters are kept by such planters in a very filthy condition; by the time the job ends the straw used for sleeping is literally converted into dung, for it is never changed: this becomes a breeding ground of infection. Typhus has had to be diagnosed immediately in the case of four or five patients brought in from one and the same plantation.' In the opinion of this doctor, 'most of the syphilis cases come from the red-beet plantations.' Mr. Feinberg rightly asserts that 'work on the plantations, which is no less injurious to the workers themselves and to the surrounding population than work in the factories, has particularly disastrous consequences, because large numbers of women and juveniles are engaged in it, and because the workers here are without the most elementary protection from society and the State'; in view of this, the author wholly supports the opinion expressed by Dr. Romanenko at the Seventh Congress of Doctors of Kharkov Gubernia that 'in issuing *compulsory* regulations, *consideration* must also be given *to the conditions of the workers on the beet plantations*. These workers lack the most essential things; they live for months under the open sky and eat from a common bowl.'" Thus, the growth of beet cultivation has enormously increased the demand for rural workers, converting the neighbouring peasantry into a rural proletariat. The increase in the number of rural workers has been but slightly checked by the inconsiderable drop in the number of workers engaged in the beet-sugar industry proper.

POTATO-STARCH PRODUCTION

From branches of technical production conducted exclusively on landlord farms let us pass to such as are more or less within the reach of the peasantry. These include, primarily, the processing of potatoes (partly also wheat and other cereals) into starch and treacle. Starch production has developed with particular rapidity in the post-Reform period owing to the enormous growth of the textile industry, which raises a demand for starch. The area covered by this branch of production is mainly the non-black-earth, the industrial, and, partly, the northern black-earth gubernias.

The *Historico-Statistical Survey* estimates that in the middle of the 60s, there were about 60 establishments with an output valued at about 270,000 rubles, while in 1880 there were 224 establishments with an output valued at 1,317,000 rubles. In 1890, according to the *Directory of Factories and Works* there were 192 establishments employing 3,418 workers, with an output valued at 1,760,000 rubles.

"In the past 25 years," we read in the *Historico-Statistical Survey*, "the number of establishments engaged in starch production has increased 4 1/2 times and the total output 10 3/4 times; nevertheless, this productivity is far from covering the demand for starch", as evidenced by the increased starch imports from abroad. Analysing the data for each gubernia, the *Historico-Statistical Survey* reaches the conclusion that our production of potato starch (unlike that of wheat-starch) is of an agricultural character, being concentrated in the hands of peasants and landlords. "Showing promise of extensive development" in the future, "it is even now furnishing our rural population with considerable advantages".

We shall see in a moment who enjoys these advantages. But first let us note that two processes must be distinguished in the development of starch production: on the one hand, the appearance of new small factories and the growth of peasant production, and on the other, the concentration of production in large steam-powered factories. For instance, in 1890 there were 77 steam-powered factories, with 52 per cent of the total number of workers and 60 per cent of the total output concentrated in them. Of these works only 11 were established before 1870, 17 in the 70s, 45 in the 80s, and 2 in 1890.

To acquaint ourselves with the economy of peasant starch production, let us turn to local investigations. In Moscow Gubernia, in 1880-81, 43 villages in 4 uyezds engaged in starch making. The number of establishments was estimated at 130, employing 780 workers and having an output valued at not less than 137,000 rubles. The industry spread mainly after the Reform; its technique gradually improved and larger establishments were formed requiring more fixed capital and showing a higher productivity of labour. Hand graters were replaced by improved ones, then horse power appeared, and finally the *drum* was introduced, considerably improving and cheapening production. Here are data we have compiled from a house-to-house census of "handicraftsmen," according to size of establishment: Thus we have here small capitalist establishments in which, as production expands, the employment of hired labour increases and the productivity of labour rises. These establishments bring the peasant bourgeoisie considerable profit, and also improve agricultural technique. But the situation of the workers in these workshops is very unsatisfactory, owing to the extremely insanitary working conditions and the long working day.

The peasants who own "grating" establishments farm under very favourable conditions. The planting of potatoes (on allotment, and chiefly on rented land) yields a considerably larger income than the planting of rye or oats. To enlarge their business the workshop owners rent a considerable amount of allotment land from the poor peasants. For example, in the village of Tsybino (Bronnitsy Uyezd), 18 owners of starch workshops (out of 105 peasant families in the village) rent allotments from peasants who have left in search of employment, and also from horseless peasants, thus adding to their own 61 allotments 133 more, which they have rented; concentrated in their hands are a total of 194 allotments, *i.e.*, 44.5 per cent of the total number of allotments in the village.

"Exactly similar things," we read in the *Returns*, "are met with in other villages where the starch industry is more or less developed" (*loc. cit.*, 42). The owners of the starch workshops have twice as much livestock as the other peasants: they average 3.5 horses and 3.4 cows per household, as against 1.5 horses and 1.7 cows among the local peasants in general. Of the 68 workshop owners (covered by the house-to-house census) 10 own purchased land, 22 rent non-allotment land and 23 rent allotment land. In short, these are typical representatives of the peasant bourgeoisie.

Exactly analogous relations are to be found in the starch making industry in the Yuryev Uyezd, Vladimir Gubernia. Here, too, the workshop owners carry on production mainly with the aid of wage-labour (out of 128 workers in 30 workshops, 86 are hired); and here, too, the workshop owners are far above the mass of the peasantry as far as stock-breeding and agriculture are concerned; they use potato pulp as feed for their cattle. Even real capitalist farmers emerge from among the peasants. Mr. Prugavin describes the farm of a peasant who owns a starch works (valued at about 1,500 rubles) employing 12 wage-workers. This peasant grows potatoes on his own farm, which he has enlarged by renting land. The crop rotation is seven-field and includes clover. For the farm work he employs from 7 to 8 workers, hired from spring to autumn ("from end to end"). The pulp is used as cattle feed, and the owner intends to use the waste water for his fields.

Mr. V. Prugavin assures us that this works enjoys "quite exceptional conditions." Of course, in any capitalist society the rural bourgeoisie will always constitute a very small minority of the rural population, and in this sense will, if you like, be an "exception." But this term will not eliminate the fact that in the starch-making area, as in all the other commercial farming areas in Russia, a class of rural entrepreneurs is being formed, who are organising capitalist agriculture.

VEGETABLE-OIL PRODUCTION

The extraction of oil from linseed, hemp, sunflower and other seeds is also frequently an agricultural industry. One can gauge the development of vegetable-oil production in the post-Reform period from the fact that in 1864 the vegetable-oil output had an estimated value of 1,619,000 rubles, in 1879 of 6,486,000 rubles, and in 1890 of 12,232,000 rubles. In this branch of production, too, a double process of development is to be observed: on the one hand, small peasant (and sometimes also landlord) oil presses producing oil for sale are established in the villages. On the other hand, large steam-driven works develop, which concentrate production and oust the small establishments. Here we are interested solely in the agricultural processing of oil-bearing plants. "The owners of the hempseed oil presses," we read in the *Historico-Statistical Survey*, "belong to the well-to-do members of the peasantry"; they attach particular value to vegetable-oil production because it enables them to obtain excellent feed for their cattle (oil cake).

Mr. Prugavin (*loc. cit.*), noting the "extensive development of the production of linseed oil" in the Yuryev Uyezd, Vladimir Gubernia, states that the peasants derive "no little advantage" from it, that crop and stock raising is conducted on a far higher level by peasants who own oil presses than by the bulk of the peasantry and that some of the oil millers also resort to the hire of rural workers. The Perm handicraft census for 1894-95 also showed that crop raising is conducted on a much higher level by handicraft oil millers than by the bulk of the peasants (larger areas under crops, far more animals, better harvests, etc.), and that this improvement in cultivation is accompanied by the hiring of rural workers.

In the post-Reform period in Voronezh Gubernia, there has been a particular development of the commercial cultivation of sunflower seed, which is crushed for oil in local presses. The area under sunflowers in Russia in the 70s was estimated at about 80,000 dess. (*Historico-Statistical Survey*, I), and in the 80s at about 136,000 dess., of which 2/3 belonged to peasants. "Since then, however, judging by certain data, the area under this plant has considerably increased, in some places by 100 per cent and even more" (*Productive Forces*, I, 37). "In the village of Alexeyevka alone" (Biryuch Uyezd, Voronezh Gubernia), we read in the *Historico-Statistical Survey*, Vol. II, "there are more than 40 oil presses, and Alexeyevka itself, solely owing to sunflowers, has prospered and grown from a wretched little hamlet into a rich town ship, with houses and shops roofed with sheet iron" (p. 41).

How this wealth of the peasant bourgeoisie was reflected in the condition of the mass of the peasantry may be seen from the fact that in 1890, in the village of Alexeyevka, out of 2,273 families registered (13,386 persons of both sexes), 1,761 had no draught animals, 1,699 had no implements, 1,480 cultivated no land, and only 33 families did not engage in industries.

TOBACCO GROWING

In conclusion, let us make some brief observations on the development of tobacco growing. The average crop in Russia for the years 1863-1867 was 1,923,000 poods from 32,161 dess.; for 1872-1878 it was 2,783,000 poods from 46,425 dess.; for the 80s, it was 4 million poods from 50,000 dess. The number of plantations in the same periods was estimated at 75,000, 95,000 and 650,000 respectively, which evidently indicates a very considerable increase in the number of small cultivators drawn into this type of commercial farming. Tobacco growing requires a considerable number of workers.

Among the types of agricultural migration note is therefore made of migration to tobacco plantations (particularly to the outer gubernias in the South, where the cultivation of tobacco has recently expanded with exceptional rapidity).

Reference has already been made in publications to the fact that the workers on the tobacco plantations lead a very hard life.

In the *Survey of Tobacco Growing in Russia*, there are very detailed and interesting data on tobacco growing as a branch of commercial farming. Mr. V. S. Shcherbachov, describing tobacco growing in Malorossia, gives wonderfully precise information on three uyezds of Poltava Gubernia (Priluki, Lokhvitsa and Romny).

This information, gathered by the author and arranged by the Bureau of Statistics, Poltava Gubernia Zemstvo Board, covers 25,089 peasant farms in the three uyezds that grow tobacco; they have 6,844 dessiatines under tobacco and 146,774 dessiatines under cereals. We see an enormous concentration of both the tobacco and the cereal area in the hands of the capitalist farms. Less than one-eighth of the farms (3,000 out of 25,000) hold more than half the area under cereals (74,000 dess. out of 147,000), with an average of nearly 25 dess. per farm. Almost half the area under tobacco (3,200 dess. out of 6,800) belongs to these farms, the average per farm being over 1 dessiatine, whereas for all the other groups the area under tobacco does not exceed one-to two-tenths of a dessiatine per household.

The owner of 7 dessiatines under tobacco must there fore have at least 14 workers; in other words, he must undoubtedly base his farm on wage-labour. Some grades of tobacco require not two but three seasonal workers per dessiatine, and day labourers in addition. In a word, we see quite clearly that the greater the degree to which agriculture becomes commercial, the more highly developed is its capitalist organisation.

The preponderance of small and tiny farms among the tobacco growers (11,997 farms out of 25,089 have *up to one-tenth* of a dessiatine planted) does not in the least refute the fact of the capitalist organisation of this branch of commercial agriculture; for this mass of tiny farms accounts for an insignificant share of the output (11,997, *i.e.*, nearly half the farms, have in all 522 dess. out of 6,844, or less than one-tenth). Nor do "average" figures, to which people so often confine themselves, provide a picture of the real situation (the average per farm is a little over 114 dessiatine under tobacco).

In some uyezds the development of capitalist agriculture and the concentration of production are still more marked. In the Lokhvitsa Uyezd, for example, 229 farms out of 5,957 each have 20 dessiatines and more under cereals.

Their owners have 22,799 dess. under cereals out of a total of 44,751, *i.e.*, more than half. Each farmer has about 100 dess. under crops. Of the land under tobacco they have 1,126 dess. out of 2,003 dess. And if the farms are grouped according to area under tobacco, we have in this uyezd 132 farmers out of 5,957 with two and more dessiatines under tobacco. These 132 farmers have 1,441 dess. under tobacco out of 2,003, *i.e.*, 72 per cent and more than *ten* dessiatines under tobacco per farm. At the other extreme of the same Lokhvitsa Uyezd we have 4,360 farms (out of 5,957) having up to one-tenth of a dessiatine each under tobacco, and altogether 133 dessiatines out of 2,003, *i.e.*, 6 per cent. It goes without saying that the capitalist organisation of *production* is accompanied here by a very considerable development of *merchant's* capital and by all sorts of exploitation outside the sphere of production. The small tobacco growers have no drying sheds, are unable to give their tobacco time to ferment and to sell it (in 3 to 6 weeks) as a finished product. They sell the unfinished product at *half the price* to buyers-up, who very often plant tobacco themselves on rented land. The buyers-up "squeeze the small planters in every way" (p. 31 of cited publication). Commercial agriculture is commercial capitalist production: this relation can be clearly traced (if only one is able to select the proper methods) in this branch of agriculture too.

AGRICULTURAL PROTECTION AND SUPPORT

A major impediment to growth in agricultural trade remains to be the high level of protection and support to agriculture in the developed countries. The Uruguay Round initiated the process of opening up new opportunities for export diversification in agriculture, through, *inter alia*; tariffication and across-the-board reductions in most-favoured nation (MFN) tariffs on agricultural products; reduction of trade-distorting support; and the strengthening of trade rules, particularly those on sanitary and phytosanitary measures and technical barriers to trade.

Despite these changes, little improvement has been achieved and market access problems persist particularly for processed agricultural products.

Agricultural Tariffs: Agricultural tariffs remain high and complex. Tariffs in the developed countries are generally lower on average, but there are many tariff peaks, particularly on processed products and temperate-zone products such as horticulture, sugar, cereals, dairy products and meat. These tariffs are often complex with several different *ad valorem*, specific and seasonal rates applying to the same products.

Bound tariffs are also high in many developing countries, although their applied rates are generally lower. Recent estimates show that nominal protection rates in the OECD countries have declined somewhat but remain high on the whole. While tariff rate quotas (TRQs) have created some new trading opportunities, consistently low and declining fill rates indicate that there are difficulties in their implementation.

Tariff Escalation: The existence of tariff escalation (tariffs which escalated with the degree of processing) in agricultural markets is regarded as one of the major factors that hinder export growth and diversification in the exporting countries. Tariff escalation prevails in a large number of agricultural commodity chains. Of 17 commodity chains examined by the Secretariat, 12 suffer from tariff escalation, mostly at the first stage of processing. In developed countries, escalation is particularly pronounced in coffee, cocoa, oilseeds, fruit and hides and skins. In developing countries, however, applied tariffs are often much lower than bound rates and in almost all cases they also show relatively high escalation.

Tariff escalation has been raised as one of the important market access issues in the current WTO negotiations on agriculture. Country proposals addressing tariff escalation are suggesting the adoption of a harmonising reduction formula that reduces higher tariffs by greater amounts, including tariff peaks, and eliminates tariff escalation. A recent FAO study shows that while the adoption of a harmonising formula for reducing tariffs such as the Swiss formula appears to be a better option for reducing tariff escalation than linear methods, it is unlikely that this formula will achieve substantial reduction in tariff escalation in all commodities of export interest to low income agricultural commodity-dependent exporters, particularly tropical products, unless it involves some additional criteria for capping tariff wedges along the processing chain.

Support to Production and Exports: Subsidized production and exports affect both domestic and international markets and exert a negative influence on the diversification of production and exports from commodity exporting countries. Loss of competitiveness caused by subsidized agriculture discourages investments in agriculture and local processing in non-subsidizing countries. Under the WTO Agreement on Agriculture (AoA), WTO members undertook to reduce both export subsidies and the trade-distorting domestic support. Export subsidies have been reduced somewhat on several products since 1995, but they remain high, particularly for meat and dairy products. In domestic support, some WTO members have shifted to less distorting support measures, but overall support to agriculture remains high.

Estimates show that total transfers to agriculture in the OECD countries amounted to US$311 billion in 2001, compared with US$298 billion in 1986-88. The support to producers in these countries as a share of the value of farm output averaged 31 per cent for agriculture as a whole and exceeded 40 per cent in such products as rice, wheat, sugar, beef and sheep meat. Such support is believed to have stimulated overproduction creating a downward pressure on world prices, thus limiting trading opportunities for non-subsidizing exporters.

International Technical Standards. A major challenge faced by many exporters of agricultural products, particularly in low income countries, is to meet the sanitary, phytosanitary and technical requirements of importing countries. An additional challenge is faced by these countries when new standards are introduced on risk assessment grounds that are stricter than those currently recognized by international standard-setting bodies. Rising consumer concerns in the affluent countries over food quality compound the difficulty of these countries in meeting ever higher standards. The transition to risk-based standard setting has required major changes in legislative, regulatory and administrative practices in most countries all of which have implied significant costs.

The concerns of low income countries in this regard are three-fold: first that they usually lack the technical and financial capacity to fully participate in activities of international standard-setting bodies such as the Codex Alimentarius Commission (CAC) and the International Plant Protection Convention (IPPC). Second, they lack the supply-side capacity to meet the increasingly strict standards being adopted by the developed countries. Third, there are concerns about the scope for the discriminatory use of these measures. A recent FAO study provided a broad indication of this problem based on data on rejections following border inspection in export markets.

The import detention list for the United States for the period 1996-97 shows that the majority of detentions and rejections of products coming from developing countries, particularly those of Sub-Saharan Africa, are related to microbiological contaminants and filth rather than to strictly technical considerations. This illustrates the considerable problems that low income countries have in meeting basic food hygiene requirements let alone requirements for which more sophisticated monitoring and testing, and therefore more costly procedures are required, such as limits on pesticide residues and heavy metals.

PREFERENTIAL MARKET ACCESS

Despite the current constraints on agricultural trade emanating from protection and support in the developed country markets, preferential trade arrangements involving developing countries have provided some opportunities for these countries to expand and diversify their agricultural exports. Three major forms of trade preferences can be distinguished: the Generalized System of Preferences (GSP), special preferential regimes for groups of developing countries (such as Lome/Cotonou, AGOA, or the Caribbean Basin Initiative), and regional free-trade areas between developed and developing countries. However, the last form differs from the others in that it involves reciprocal rather than unilateral trade preferences.

MFN duties are still high for many agricultural products and preference schemes are offering substantial price advantages to some exporters for numerous eligible products. This is specially the case, on a wide scale, for least developed countries, as well as under special preferential arrangements of the EU and the United States for most African, Caribbean, Central American, Andean and Pacific countries.

While many developing countries have been accorded preferences for a large number of products, exceptions to these preferences often relate to agricultural products. For example, the European Union's initiative on "Everything but Arms" (EBA) offers free market access to LDC products, with less than 5 per cent of pre-EBA exports left facing a tariff barrier. A recent study by UNCTAD, however, shows that the impact of this initiative will be a relatively small increase, in the medium term, in exports from the LDCs, as 70 per cent of the potential positive trade effects would have come from free access for sugar, rice and beef, which has been deferred until 2006.

The African Growth and Opportunity Act has also started to open up markets in the United States for African countries. Although trade preferences are of significant benefits to a number of countries, in general they have been, to a large extent, underutilised. Part of the reason for this underutilisation is due to, in many cases, lack of security of access, insufficient product coverage, excessively stringent rules of origin and often a lack of awareness of the preferences available. These factors are in addition to supply-side constraints faced by some countries, in particular the least developed countries

DISTRIBUTION OF GAINS

Government barriers to trade are important, but as they are reduced other factors impeding market entry, particularly those that spring from restrictive business practices, come to the forefront.

The agricultural commodity chains, particularly those of high value crops and processed products, are increasingly dominated by a few multinational enterprises (MNEs) and distribution companies. This dominance of large companies in the world agricultural commodity markets is not only through their actions on international markets, but also through their direct and increasing influence on what is produced, and how. While these changes are opening up unprecedented opportunities for some producers and exporters, benefiting from these developments and avoiding their negative impacts requires much greater business skills than before on the part of Governments and small-scale entrepreneurs, particularly in low-income countries.

Changes in agricultural commodity chains, as a result of the increasing dominance of large companies, can be reviewed at three levels, namely, the international markets, exporting developing countries, and the retail markets of importing countries.

At the international level, agricultural trade continues to be concentrated in a few vertically integrated MNEs. In 1996, for example, five MNEs accounted for about 50 per cent of world trade in green coffee, up from about 37 per cent in 1980; and four coffee roasters accounted for 50 per cent of the roasted coffee market. In cocoa, the number of trading houses in London has been reduced from 30 in 1980 to around 10 in 1999. Similarly, the six largest chocolate manufacturers account for 50 per cent of world chocolate sales. For vegetable oils, following a series of mergers and acquisitions in the 1990s, a small number of vertically integrated MNEs now dominate the production, distribution and international trade of both oilseeds and oils. In grains, a few big companies are transforming themselves, through mergers and consolidations, into vertically integrated businesses in trading, storage, processing and milling. In exporting developing countries, following the deregulation and the disappearance of many marketing boards, large companies with warehousing and shipping facilities in the producing countries are able to exploit their financial and logistical advantages, even buying the produce directly from the farmer. Intensified competition favours those with access to cheaper finance and good logistics. Being big provides advantages on both accounts.

At the retail level, supermarkets have also being growing rapidly in both developed and developing countries. For example, there has been rapid consolidation and multinationalisation of the supermarket sector in Latin America, where the share of supermarkets in food retailing went from 15-20 per cent in 1990 to 60 per cent in 2000.

Given their dominance in the market, supermarkets have a significant impact on production, distribution and trade. To simplify operations, most of these supermarkets are increasingly cutting the number of their suppliers, and as a result those left out will have to sell in a shrinking and low-priced market or find a niche market. In addition, supermarkets have significant impact on production requirements in terms of quality and food safety, product variety and delivery. The Food Business Forum (CIES), through its association of the world's 200 largest supermarkets and their suppliers (with a combined sales of US$2 800 billion), decided earlier this year on new mandatory standards for their suppliers. While this move may help in reducing the number of different standards, it may overshadow whatever Governments decide in the context of the WTO agreements.

The stringent private standards set by supermarkets are very often a reflection of consumers' concerns. Consumers of agricultural commodities are increasingly choosy with respect to the quality of the commodities they buy, the timeliness of delivery, origin and traceability, and the environmental and social conditions under which they have been produced. The use of specialised channels such as fair-trade labelling and other socially or environmentally responsible market practices, and taking advantage of niche market such as that for organic food, may help in facilitating market entry and can improve earnings of producers. In many countries, consumers are willing to pay a price premium (of more than 20 per cent) for organic foods, particularly fruit and vegetables.

There is evidence that the increasing market concentration along agricultural commodity chains has been associated with a declining share of producing countries in the revenues generated from the sale of these products in global markets. Recent studies show a widening gap between world commodity prices and consumer prices in industrialized countries. Growers' price represents very low shares of the final price of the product, ranging from 4-8 per cent for raw cotton and tobacco to 11– 24 per cent for jute and coffee. Similar patterns were also observed in fresh fruit and vegetables. The International Coffee Organisation (ICO) has reported that in the early 1990s export earnings by coffee producing countries were about US$10-12 billion and the value of retail sales of coffee, largely in the developed countries, was about US$30 billion. However, in coffee year 2000/01, producing countries received only US$5.8 billion of the value of retail sales of more than US$60 billion.

Studies on commodity chains involving exports of fresh vegetables and fruit from Africa to the UK have shown that returns to the participant in the market are highly concentrated at the end of the chain in the importing countries.

2

Dairy Farming

Dairying is an important source of subsidiary income to small/marginal farmers and agricultural labourers. The manure from animals provides a good source of organic matter for improving soil fertility and crop yields. The gober gas from the dung is used as fuel for domestic purposes as also for running engines for drawing water from well. The surplus fodder and agricultural by-products are gainfully utilised for feeding the animals. Almost all draught power for farm operations and transportation is supplied by bullocks. Since agriculture is mostly seasonal, there is a possibility of finding employment throughout the year for many persons through dairy farming. Thus, dairy also provides employment throughout the year. The main beneficiaries of dairy programmes are small/marginal farmers and landless labourers. A farmer can earn a gross surplus of about ₹ 12,000 per year from a unit consisting of 2 milking buffaloes. The capital investment required for purchase of 2 buffaloes is ₹ 18,223/-. Even after paying a sum of ₹ 4294/- per annum towards repayment of the loan and interest the farmer can earn a net surplus of ₹ 6000 - 9000/- approximately per year.

Dairy farming is a class of agricultural, or an animal husbandry, enterprise, for long-term production of milk, usually from dairy cows but also from goats and sheep, which may be either processed on-site or transported to a dairy factory for processing and eventual retail sale.

Most dairy farms sell the male calves born by their cows, usually for veal production, or breeding depending on quality of the bull calf, rather than raising non-milk-producing stock. Many dairy farms also grow their own feed, typically including corn, alfalfa, and hay. This is fed directly to the cows, or is stored as silage for use during the winter season. Additional dietary supplements are often added to the feed to increase quality milk production.

UNPROFITABLE OF DAIRY FARMING

Dairy farmers in India are in crisis. High input prices and declining milk prices have pushed them to the wall. But the consumer prices remain firm, showing no signs of declining. In 2009, about 20 per cent of 1,800 dairy farms in California, for instance, had shut down unable to survive at times of higher feed and transportation costs.

Similarly, in 2009, when international milk prices had dipped to a low, the European Union defied the World Trade Organisation (WTO) and reintroduced milk subsidies. It provided ₹ 3,600 crore in subsidies to its dairy farmers to offset the losses incurred.

In America, on the other hand, the number of dairy farms has come down by 61 per cent since 1992, and only 51,480 dairy farms now exist. These farms received subsidies worth ₹ 27,500 crore since 2009, which in other words means a third of its milk price is subsidised. These subsidies come under several programmes: milk income loss contract payment; market loss assistance; milk income loss transitional payment; dairy economic loss assistance programme; milk marketing fees; dairy disaster assistance; and dairy indemnity.

Let me also dispel another commonly held notion. In an era of market economy, it is generally believed that American/European farmers are dependent entirely on the private supply chain. It's not so. After March 2009, EU had restarted buying surplus butter and milk from dairy farmers at an intervention price of Euro 2,218 and Euro 1,698 per tonne, respectively. In America, milk price support programme ensures that the government buys any surplus amount of cheese, butter and non-fat dry milk at a minimum price. In addition, since 2002, it has introduced a programme to distribute cash subsidies to milk producers when prices fall below a set limit. I see no reason why state governments cannot provide such subsidy support to its dairy farmers when prices fall.

Therefore, to begin with, state governments should consider subsidising the cooperatives to raise the milk procurement price to at least ₹ 25 per litre. Secondly, it should also reduce the bank interest on loans taken from existing 12.5 per cent to 7 per cent.

VACUUM BUCKET MILKING

The first milking machines were an extension of the traditional milking pail. The early milker device fit on top of a regular milk pail and sat on the floor under the cow. Following each cow being milked, the bucket would be dumped into a holding tank. This developed into the Surge hanging milker.

Prior to milking a cow, a large wide leather strap called a surcingle was put around the cow, across the cow's lower back. The milker device and collection tank hung underneath the cow from the strap. This innovation allowed the cow to move around naturally during the milking process rather than having to stand perfectly still over a bucket on the floor.

With the availability of electric power and suction milking machines, the production levels that were possible in stanchion barns increased but the scale of the operations continued to be limited by the labour intensive nature of the milking process. Attaching and removing milking machines involved repeated heavy lifting of the machinery and its contents several times per cow and the pouring of the milk into milk cans. As a result, it was rare to find single-farmer operations of more than 50 head of cattle.

STEP-SAVER MILK TRANSPORT

As herd size began to increase, the bucket milker system became laborious. A vacuum milk-transport system known as the Step-Saver was developed to transport milk to the storage tank. The system used a long vacuum hose coiled around a receiver cart, and connected to a vacuum-breaker device in the milkhouse, allowing farmers to milk many cows without the necessity of walking increasingly longer distances carrying heavy buckets of milk.

MILKING PIPELINE

The next innovation in automatic milking was the milk pipeline. This uses a permanent milk-return pipe and a second vacuum pipe that encircles the barn or milking parlor above the rows of cows, with quick-seal entry ports above each cow. By eliminating the need for the milk container, the milking device shrank in size and weight to the point where it could hang under the cow, held up only by the sucking force of the milker nipples on the cow's udder.

The milk is pulled up into the milk-return pipe by the vacuum system, and then flows by gravity to the milkhouse vacuum-breaker that puts the milk in the storage tank. The pipeline system greatly reduced the physical labour of milking since the farmer no longer needed to carry around huge heavy buckets of milk from each cow.

The pipeline allowed barn length to keep increasing and expanding, but after a point farmers started to milk the cows in large groups, filling the barn with one-half to one-third of the herd, milking the animals, and then emptying and refilling the barn. As herd sizes continued to increase, this evolved into the more efficient milking parlor.

MILKING PARLORS

Innovation in milking focused on mechanizing the milking parlor to maximize throughput of cows per operator which streamlined the milking process to permit cows to be milked as if on an assembly line, and to reduce physical stresses on the farmer by putting the cows on a platform slightly above the person milking the cows to eliminate having to constantly bend over. Many older and smaller farms still have tie-stall or stanchion barns, but worldwide a majority of commercial farms have parlors. The milking parlor allowed a concentration of money into a small area, so that more technical monitoring and measuring equipment could be devoted to each milking station in the parlor. Rather than simply milking into a common pipeline for example, the parlor can be equipped with fixed measurement systems that monitor milk volume and record milking statistics for each animal. Tags on the animals allow the parlor system to automatically identify each animal as it enters the parlor.

RECESSED PARLORS

More modern farms use recessed parlors, where the milker stands in a recess such that his arms are at the level of the cow's udder. Recessed parlors can be herringbone, where the cows stand in two angled rows either side of the recess and the milker accesses the udder from the side, parallel, where the cows stand side-by-side and the milker accesses the udder from the rear or, more recently, rotary, where the cows are on a raised circular platform, facing the centre of the circle, and the platform rotates while the milker stands in one place and accesses the udder from the rear. There are many other styles of milking parlors which are less common.

MILK PRESERVATION METHODS

Cool temperature has been the main method by which milk freshness has been extended. When windmills and well pumps were invented, one of its first uses on the farm besides providing water for animals was for cooling milk, to extend the storage life before being transported to the town market.

The naturally cold underground water would be continuously pumped into a tub or other containers of milk set in the tub to cool after milking. This method of milk cooling was extremely popular before the arrival of electricity and refrigeration.

REFRIGERATION

When refrigeration first arrived the equipment was initially used to cool cans of milk, which were filled by hand milking. These cans were placed into a cooled water bath to remove heat and keep them cool until they were able to be transported to a collection facility. As more automated methods were developed for harvesting milk, hand milking was replaced and, as a result, the milk can was replaced by a bulk milk cooler. 'Ice banks' were the first type of bulk milk cooler.

This was a double wall vessel with evaporator coils and water located between the walls at the bottom and sides of the tank. A small refrigeration compressor was used to remove heat from the evaporator coils. Ice eventually builds up around the coils, until it reaches a thickness of about three inches surrounding each pipe, and the cooling system shuts off. When the milking operation starts, only the milk agitator and the water circulation pump, which flows water across the ice and the steel walls of the tank, are needed to reduce the incoming milk to a temperature below 40 degrees.

This cooling method worked well for smaller dairies, however was fairly inefficient and was unable to meet the increasingly higher cooling demand of larger milking parlors. In the mid 1950's direct expansion refrigeration was first applied directly to the bulk milk cooler. This type of cooling utilizes an evaporator built directly into the inner wall of the storage tank to remove heat from the milk. Direct expansion is able to cool milk at a much faster rate than early ice bank type coolers and is still the primary method for bulk tank cooling today on small to medium sized operations.

Another device which has contributed significantly to milk quality is the plate heat exchanger. This device utilizes a number of specially designed stainless steel plates with small spaces between them. Milk is passed between every other set of plates with water being passed between the balance of the plates to remove heat from the milk.

This method of cooling can remove large amounts of heat from the milk in a very short time, thus drastically slowing bacteria growth and thereby improving milk quality. Ground water is the most common source of cooling medium for this device. Dairy cows consume approximately 3 gallons of water for every gallon of milk production and prefer to drink slightly warm water as opposed to cold ground water. For this reason, PHE's can result in drastically improved milk quality, reduced operating costs for the dairymen by reducing the refrigeration load on his bulk milk cooler, and increased milk production by supplying the cows with a source of fresh warm water.

Plate heat exchangers have also evolved as a result of the increase of dairy farm herd sizes in the US. As a dairyman increases the size of his herd, he must also increase the capacity of his milking parlor in order to harvest the additional milk.

This increase in parlor sizes has resulted in tremendous increases in milk throughput and cooling demand. Today's larger farms produce milk at a rate which direct expansion refrigeration systems on bulk milk coolers cannot cool in a timely manner.

PHE's are typically utilized in this instance to rapidly cool the milk to the desired temperature before it reaches the bulk milk tank. Typically, ground water is still utilized to provide some initial cooling to bring the milk to between 55 and 70 degrees F. A second part of the PHE is added to remove the remaining heat with a mixture of chilled pure water and propylene glycol. These chiller systems can be made to incorporate large evaporator surface areas and high chilled water flow rates to cool high flow rates of milk.

MILKING OPERATION

Milking machines are held in place automatically by a vacuum system that draws the ambient air pressure down from 15 to 21 pounds per square inch of vacuum. The vacuum is also used to lift milk vertically through small diametre hoses, into the receiving can. A milk lift pump draws the milk from the receiving can through large diametre stainless steel piping, through the plate cooler, then into a refrigerated bulk tank. Milk is extracted from the cow's udder by flexible rubber sheaths known as liners or inflations that are surrounded by a rigid air chamber. A pulsating flow of ambient air and vacuum is applied to the inflation's air chamber during the milking process.

When ambient air is allowed to enter the chamber, the vacuum inside the inflation causes the inflation to collapse around the cow's teat, squeezing the milk out of teat in a similar fashion as a baby calf's mouth massaging the teat. When the vacuum is reapplied in the chamber the flexible rubber inflation relaxes and opens up, preparing for the next squeezing cycle.

It takes the average cow three to five minutes to give her milk. Some cows are faster or slower. Slow-milking cows may take up to fifteen minutes to let down all their milk. Milking speed is only minorly related to the quantity of milk the cow produces — milking speed is a separate factor from milk quantity; milk quantity is not determinative of milking speed. Because most milkers milk cattle in groups, the milker can only process a group of cows at the speed of the slowest-milking cow. For this reason, many farmers will cull slow-milking cows.

The extracted milk passes through a strainer and plate heat exchangers before entering the tank, where it can be stored safely for a few days at approximately 42 °F. At pre-arranged times, a milk truck arrives and pumps the milk from the tank for transport to a dairy factory where it will be pasteurized and processed into many products.

AUTOMATIC MILKING

They have been the topic of discussion for some time now, but as they become increasingly popular, Charlotte Johnston, TheCattleSite junior editor, looks at just how effective automatic milking machines are and evaluates their benefits. A common myth regarding automatic milking machines is that they disengage the herdsman from the animals, and so neglect herd health. If this were true, the new Lely Astronaut A3 Next would not have won the 2009 Royal Association of British Dairy Farmers' Livestock and

Machinery Award at this year's Dairy Event. The award was presented on the grounds of economics, the welfare and close management of milking animals, the reduction of drudgery for family and employed staff, flexibility since it applies to family run and larger units, and the longer-term needs of the dairy industry. TheCattleSite spoke to four dairy farmers around the world to find out how automatic milking equipment changed their lives. Click on *Tim Gibson*, *Doug Heintz*, *John Wolf* and *Max Warren* to read about their experiences.

HOW DOES IT WORK?

Very simply, a laser control will identify the location of the teats and a robust robot arm will attach cups individually to each teat from underneath the cow. Swift teat cup attachment is guaranteed even when milking cows with teats missing. Should a cup be kicked off, it is reconnected quickly.

COW ID SYSTEM AND COMPUTER MANAGEMENT SOFTWARE

Each cow has an electronic tag, which is read when she enters the machine. This allows the computer system to identify each cow. Management through the computer tells the stockman specific data for each individual cow. The MQC sensor system checks the milk colour, conductivity and flow to ensure maximum control of milk quality. Comparing individual cow data to the average allows problems to be identified.

Milk identified with blood is separated and either discarded or diverted into another holding place. Furthermore, if a cow is being treated with medicine, it is possible for her to remain in the milking herd. This does require strict attention to detail - as accurate information must be input into the computer system as to what the cow has been treated with, how long the withdrawal period is etc.

When the cow comes in to be milked next, her milk will be automatically diverted into a separate holding area or disposed of. The same can be done for colostrum, for example, when a cow has recently calved as it may be beneficial to retain the colostrum for a couple of days. Entering all information on the system will allow this milk to be stored separately and after two days, her milk will return to the bulk tank. By checking the computer twice a day, the stockman can identify cows that are overdue milking or can treat illnesses identified immediately and the accessibility to such detailed data can allow accurate management decisions to be made.

DYNAMIC FEEDING

Dynamic feeding looks at an individual cow's feed requirements in relation to her yield. The programme takes into account milk yield, the cost of feed and the milk price received to work out the most economical feed rate for each individual cow. For Max Warren in Australia, this is particularly useful in a drought situation, where extra feed may have to be bought, but when milk prices are particularly low. On average in the UK, for every litre of milk produced per cow, 0.3kg of dry matter is consumed, says Tim Gibson. This is standard regardless of whether they are high yielders. He believes he has saved one kilo of dry matter per cow per day, which with 140 cows equates to more than 51 tonnes of dry matter a year – in other words, a lot of money!

John Wolf and Doug Heintz from the US say that they now have the ability to feed a higher concentrate diet to higher yielding cows and they believe they have reduced feed costs. The dynamic feeding programme allows optimum energy conversion efficiency. This unique and efficient system would not be possible without automatic milking machines.

LABOUR SAVINGS

All four farmers have cut back labour units. In Australia, Mr Warren believes he would require double the number of staff to milk 300 cows on a rotary system. Mr Wolf is saving 200 hours of labour a week, whilst Mr Gibson and Mr Heintz have cut back by almost one labour unit each. Adding up the cost of current labour, it is not hard to see how much money can be saved. Lely estimate that for every one machine milking 70 cows, half a labour unit is saved.

Two robots would milk 140 cows and save three-quarters of a labour unit, whilst three machines would milk 200 plus cows and save one full labour unit. "The biggest advantage by far is the labour saving. The flexibility has changed my life allowing me to spend more time with my family," says Mr Heintz.

BUILDING SPACE

Mr Gibson recommends that an area of approximately 15ft by 12ft is needed. The machines, however, are only 15ft by 6ft; the extra space takes into account a clean area behind the machine. With so little space required per machine, customers have the opportunity to increase indoor housing.

CONTROLLED GRAZING

Both Mr Wolf and Mr Heintz keep cows housed all year round; only dry cows and heifers may be put to grass. Mr Gibson allows his cows to graze outside during the summer with controlled grazing. Unlike the other farmers, Mr Warren's cows are grazed outdoors all year around on a three-way grazing system. Mr Warren operated a two-way grazing system, where cows moved paddocks every 12 hours. He found with this system that cows were hanging around the parlour and so he changed the process to a three-way grazing system. Now, cows move fields every eight hours, each time passing through the parlour. Cows are fed a wagon mix in the shed and fed concentrates when milked. All pasture is irrigated.

TRAINING HERD TO USE MACHINE

Mr Wolf said that it takes about three to four days to train heifers to use the machines. However, when the machines were introduced, it took about three weeks before things were running smoothly.

He believes that there will be about 0.5 per cent of each herd that does not adapt. Mr Heintz says that he spent three weeks pushing cows up to the machines, but it probably took about three months until the machines were working at full efficiency.

Over in Australia, Mr Warren says that it took two to three weeks to train the herd to walk into the robots but he felt it was probably twelve months until he was fully trained to use the system, the technology including computers was completely new to him!

HERD BENEFITS

The Lely Astronaut has a refined udder preparation procedure. Teats are cleaned individually and sprayed. The machine also automatically washes three times a day as well as washing in between each cow.

At the Robotic Dairy in Australia, cows are lasting six lactations. Further improvements in herd health can be seen by Mr Gibson reporting an increase in lifetime lactations per cow. Mr Wolf and Mr Warren reported improvements in foot health and reproductive performance.

Mr Heintz, who believes he had high standards of animal health prior to the installation of the machines, has reported no significant improvements in herd health but said that previous levels had been easily maintained. He has also only been using the machines for just under a year.

Walking around Mr Gibson's yard, it was apparent that the cows seemed dramatically calmer and at ease than in an conventional dairy unit. In fact, all farmers mentioned that cows were a lot less stressed, which in turn they believed led to better health and increased yields.

Increased yields seemed apparent across the board but the improvement varied between farms. In the UK, Mr Gibson reported an increase of 1,000 litres per cow over the last eight years, whereas Mr Heintz reported an increase of 10 to 15 pounds per day over the last ten months, which equates to 4562.5 pounds of milk per year or about an extra 2,000 litres per year.

SERVICE AND INSTALLATION

The same system operates in the US and the UK with Lely providing onsite support for the first 2-3 days of training cows. During the first two to three weeks, representatives will call in three or four times a week to help with the start-up. Chad Huyser, Director of Sales and Operations for Lely US, says that farm management support staff will visit a couple of times during the first few months to do a thorough review of the farm management process, which includes making slight adjustments to feed rations and robot performance settings. Mr Gibson, who completed farm management training in Holland with Lely, says that he goes onto clients' farms to help customers adapt to the new style of management and prevent problems.

Lely also offer web training events with the farm management support staff, providing information to farmers on how to use the T4C, which is the onboard management software on the robots. Mr Huyser says that one of the areas that Lely focuses on globally is providing on-going support to all their installations. Speaking with Mr Warren, he feels that ongoing technical support is essential, and is more than pleased with the service Lely continues to provide. He thoroughly enjoys working with the company and together, they have made a number of changes to adapt to Australian grazing conditions. The new Lely Astronaut A3 robotic milking system requires servicing three times a year. Mr Gibson says the older models require servicing every eight weeks. He says that information from each service is sent back to a database at the headquarters, which allows engineers to access a vast amount of data to refer to in the case of a fault.

Due to the relative newness of the machines, there is no guarantee how long they will last. Mr Gibson believes that the oldest one in place is approximately 17 years old and still going strong. There have been no reports of machines wearing out to date. Lely estimates that products with the correct service and care will last at least 30 years. Machines can be constantly updated with new software, which comes at no extra cost to the customer.

Some people may be under the impression that robotic milking systems require someone to be on call 24/7 in case of a breakdown. This is not technically true. If there is an error in the system, an alarm system rings a telephone to tell whoever is on duty. During the day, this is not a problem but during the night, it could be more than a nuisance.

Not one of the farmers complained about this. During the evenings, Mr Gibson and Mr Warren said their machines may go off once a week. With seven machines, Mr Wolf is called three or four times a week. Mr Wolf stressed that the system can be manipulated and managed to reduce these calls through only five minutes of maintenance per robot per day.

HOW MUCH DOES IT COST?

The latest model, Lely Astronaut A3 Next, is around £110,000 for one robot. Opportunities for second-hand robots exist: all robots are refurbished at the Lely factory and come with a one-year warranty. Mr Gibson says that they often become available due to clients upgrading or retiring and cost approximately 75 per cent of the new price.

FARM BENEFITS

Mr Wolf explained that the introduction of robotic milking machines has allowed greater flexibility across the farm - particularly with regard to time schedules. "Having the robots in place means that I can now finish what I was doing in the field without rushing back to the parlour to milk the cows," he said. Mr Wolf also stressed that the working environment has significantly improved with cleaner areas for both cows and employees. But by far the biggest benefit for him was the decreased costs of production.

When asked about the biggest benefit of installing a robotic milking system, Mr Warren said the fact he was still farming. "Without robots, I would have stopped milking years ago. I really enjoy working with automatic milking machines and Lely. I love the technology that is always improving. It makes my work so much more enjoyable and interesting."

DAIRY PRODUCT

Dairy products are generally defined as foods produced from cow's or domestic buffalo's milk. They are usually high-energy-yielding food products. A production plant for such processing is called a dairy or a dairy factory. Raw milk for processing mostly comes from cows and to a lesser amount from domestic buffalos, but occasionally from other mammals such as goats, sheep, yaks, or horses. Dairy products are commonly found in European, Middle Eastern and Indian cuisine, whereas they are almost unknown in East Asian cuisine.

TYPES OF DAIRY PRODUCTS

- Milk after optional homogenization, pasteurization, in several grades after standardization of the fat level, and possible addition of bacteria Streptococcus lactis and Leuconostoc citrovorum
  - Crème fraîche, slightly fermented cream
    a. Smetana, Central and Eastern European variety of sour cream
    b. Clotted cream, thick spoonable cream made by heating
  - Cultured buttermilk, fermented concentrated (water removed) milk using the same bacteria as sour cream
  - Kefir, fermented milk resembling buttermilk but based on different yeast and bacteria culture
  - Kumis/Airag, slightly fermented mares' milk popular in Central Asia
  - Milk powder (or powdered milk), produced by removing the water from milk
    a. Whole milk products

b. Buttermilk products
c. Skim milk
d. Whey products
e. Ice Cream
f. High milk-fat and nutritional products (for infant formulas)
g. Cultured and confectionery products

- Condensed milk, milk which has been concentrated by evaporation, often with sugar added for longer life in an opened can
- Evaporated milk, (less concentrated than condensed) milk without added sugar
- Ricotta cheese, milk heated and reduced in volume, known in Indian cuisine as Khoa
- Infant formula, dried milk powder with specific additives for feeding human infants
- Baked milk, a variety of boiled milk that has been particularly popular in Russia

• Butter, mostly milk fat, produced by churning cream
  - Buttermilk, the liquid left over after producing butter from cream, often dried as livestock food
  - Ghee, clarified butter, by gentle heating of butter and removal of the solid matter
  - Smen, a fermented clarified butter used in Moroccan cooking.
  - Anhydrous milkfat

• Cheese, produced by coagulating milk, separating from whey and letting it ripen, generally with bacteria and sometimes also with certain molds
  - Curds, the soft curdled part of milk (or skim milk) used to make cheese (or casein)
  - Whey, the liquid drained from curds and used for further processing or as a livestock food
  - Cottage cheese
  - Quark
  - Cream cheese, produced by the addition of cream to milk and then curdled to form a rich curd or cheese made from skim milk with cream added to the curd
  - Fromage frais

• Casein
  - Caseinates
  - Milk protein concentrates and isonates
  - Whey protein concentrates and isonates

   - Hydrolysates
   - Mineral concentrates
- Yogurt, milk fermented by Streptococcus salivarius ssp. thermophilus and Lactobacillus delbrueckii ssp. bulgaricus sometimes with additional bacteria, such as Lactobacillus acidophilus
   - Ayran
   - Lassi
- Clabber (food), milk naturally fermented to a yogurt-like state
- Gelato, slowly frozen milk and water, lesser fat than ice cream
- Ice cream, slowly frozen cream and emulsifying additives
   - Ice milk
   - Frozen custard
   - Frozen yogurt, yogurt with emulsifiers that is frozen
- Other
   - Viili
   - Kajmak
   - Filmjölk
   - Piimä
   - Vla
   - Dulce de leche

HEALTH RISKS OF CONSUMING DAIRY PRODUCTS

Most dairy products contain large amounts of saturated fat. Some dairy products may cause health issues for individuals who have a lactose intolerance and milk allergies. Some dairy products such as blue cheese may become contaminated with the fungus Aspergillus fumigatus during ripening, which can trigger asthma and other respiratory problems in susceptible individuals. Vegans and some vegetarians avoid dairy products due to a variety of ethical, dietary, environmental, political, and religious concerns.

DAIRY CATTLE

Dairy cattle (dairy cows) are cattle cows (adult females) bred for the ability to produce large quantities of milk, from which dairy products are made. Dairy cows generally are of the species Bos taurus. Historically, there was little distinction between dairy cattle and beef cattle, with the same stock often being used for both meat and milk production. Today, dairy cows are specialized and most have been bred to produce large volumes of milk, with little or no regard for their production of meat. Between 1959 and 1990, US milk production doubled while the number of dairy cows declined 40 per cent.

COW

Dairy cows may be found either in herds on dairy farms where dairy farmers own, manage, care for, and collect milk from them, or on commercial farms. Dairy cow herds range in size from small farms of fewer than five cows to large herds of about 20,000. The average dairy farmer in the United States manages about one hundred cows but this varies from an average of 800 cows in California to under 80 in the North East states. Herd sizes vary around the world depending on landholding culture and social structure. In many European countries the average herd size is well below 50. In the UK it is over 100 in New Zealand 350 and Australia 280.

To maintain high milk production, a dairy cow must be bred and produce calves. Depending on market conditions, the cow may be bred with a "dairy bull" or a "beef bull." Female calves (heifers) with dairy breeding may be kept as replacement cows for the dairy herd. If a replacement cow turns out to be a substandard producer of milk, she then goes to market and can be killed for beef.

Male calves can either be used later as a breeding bull or sold and used for veal or beef. Most dairy farmers begin breeding heifers at fifteen months of age. A cow's gestation period is approximately nine months, so most heifers give birth at around two years of age. A Californian industry report, the natural lifespan of a dairy cow is approximately 15–25 years, however dairy cows are rarely kept longer than five years prior to slaughter. Herd life is strongly correlated with production levels. Approximately 17 per cent of the US beef supply comes from cull dairy cows: cows that can no longer be seen as a economic asset to the dairy farm. These animals may be sold due to common diseases of milk cows including mastitis, lameness, or other diseases.

In India, the Hindu majority holds the cow as sacred and a motherly figure due to her capacity to give milk. Cow slaughter is legally banned in India (except in the state of Kerala, West Bengal and the seven north eastern states). Thus, spent dairy cows don't go to slaughter, but are often seen as roaming on the city streets and die of old age or disease. Some pious Hindu organizations manage "old age homes" for old dairy cows.

CALF

Market calves are generally sold at two weeks of age and bull calves may fetch a premium over heifers due to their size, either current or potential. Calves may be sold for veal, or for one of several types of beef production, depending on available local crops and markets. Such bull calves may be castrated if turnout onto pastures is envisaged, in order to render the animals less aggressive. Purebred bulls from elite cows may be put into progeny testing schemes to find out whether they might become superior sires for breeding. Such animals may become extremely valuable. Most dairy farms separate calves from their mothers within a day of birth to reduce transmission of disease and simplify management of milking cows. Studies have been done allowing calves to remain with their mothers for 1, 4, 7 or 14 days after birth. Cows whose calves were removed longer than one day after birth showed increased searching, sniffing and vocalizations.

However, calves allowed to remain with their mothers for longer periods showed weight gains at three times the rate of early removals as well as more searching behaviour and better social relationships with other calves. After separation, most young dairy calves subsist on commercial milk replacer, a feed based on dried milk powder. Milk replacer is an economical alternative to feeding whole milk because it is cheaper, can be bought at varying fat and protein percentages. A day old calf consumes around 5 liters of milk per day.

BULL

A bull calf with high genetic potential may be reared for breeding purposes. It may be kept by a dairy farm as a herd bull, to provide natural breeding for the herd cows. A bull may service up to 50 or 60 cows during a breeding season. Any more and the sperm count will decline, leading to cows "returning to service" (to be bred again). A herd bull may only stay for one season since over two years old their temperament becomes too unpredictable. Bull calves intended for breeding commonly are bred on specialized dairy breeding farms, not production farms. These farms are the major source of stocks for artificial insemination (AI).

MILK PRODUCTION LEVELS

A cow will produce large amounts of milk over her lifetime. Certain breeds produce more milk than others; however, different breeds produce within a range of around 15,000 to 25,000 lbs of milk per lactation. The average for dairy cows in the US in 2005 was 19,576 pounds. Production levels peak at around 40 to 60 days after calving. The cow is then bred. Production declines steadily afterwards, until, at about 305 days after calving, the cow is 'dried off', and milking ceases.

About sixty days later, one year after the birth of her previous calf, a cow will calve again. High production cows are more difficult to breed at a one year interval. Many farms take the view that 13 or even 14 month cycles are more appropriate for this type of cow. Dairy cows may continue to be economically productive for many lactations. Ten or more lactations are possible. The chances of problems arising which may lead to a cow being culled are however, high; the average herd life of US Holsteins is today fewer than 3 lactations. This requires more herd replacements to be reared or purchased.

Over 90 per cent of all cows are culled for 4 main reasons:

1. *Infertility*: Failure to conceive and reduced milk production. Cows are at their most fertile between 60 and 80 days after calving. Cows remaining "open" (not with calf) after this period become increasingly difficult to breed, which may be due to poor health. Failure to expel the afterbirth from a previous pregnancy, luteal cysts, or metritis, an infection of the uterus, are common causes of infertility.
2. *Mastitis*: Persistent and potentially fatal mammary gland infection, leading to high somatic cell counts and loss of production. Mastitis is recognized by a reddening and swelling of the infected quarter of the udder and the presence of whitish clots or pus in the milk. Treatment is possible with long-acting antibiotics but milk from such cows is not marketable until drug residues have left the cow's system.
3. *Lameness*: Persistent foot infection or leg problems causing infertility and loss of production. High feed levels of highly digestible carbohydrate cause acidic conditions in the cow's rumen. This leads to laminitis and subsequent lameness, leaving the cow vulnerable to other foot infections and problems which may be exacerbated by standing in feces or water soaked areas.
4. *Production*: Some animals fail to produce economic levels of milk to justify their feed costs. Production below 12 to 15 liters of milk per day are not economically viable.

Herd life is strongly correlated with production levels. Lower production cows live longer than high production cows, but may be less profitable. Cows no longer wanted for milk production are sent to slaughter. Their meat is of relatively low value and is generally used for processed meat.

REPRODUCTION

Since the 1950s, artificial insemination (AI) is used at most dairy farms; these farms may keep no bull. Advantages of using AI include its low cost and ease compared to maintaining a bull, ability to select from a large number of bulls to match the anticipated market for the resulting calves, and predictable results.

More recently, embryo transfer has been used to enable the multiplication of progeny from elite cows. Such cows are given hormone treatments to produce multiple embryos.

These are then 'flushed' from the cow's uterus. 7-12 embryos are consequently removed from these donor cows and transferred into other cows who serve as surrogate mothers. The result will be between 3 and 6 calves instead of the normal single, or rarely, twins.

HORMONE USE

Hormone treatments are given to dairy cows to increase reproduction and to increase milk production. The hormones are used to produce multiple embryos have to be administered at specific times to dairy cattle to induce ovulation.

Frequently, for economic considerations, these drugs are also used to synchronize a group of cows to ovulate simultaneously. The hormones Prostaglandin, Gonadotropin Releasing Hormone, and Progesterone are used for this purpose and sold under the brand names Lutalyse, Cystorelin, Estrumate, Factrel, Prostamate, Fertagyl. Insynch, and Ovacyst. They may be administered by injection, insertion or mixed with feed.

About 17 per cent of dairy cows in the United States are injected with Bovine somatotropin, also called recombinant bovine somatotropin (rBST), recombinant bovine growth hormone (rBGH), or artificial growth hormone. The use of this hormone increases milk production from 11 per cent-25 per cent, but also increases the likelihood of cattle developing mastitis, reduction in fertility and lameness.

The U.S. Food and Drug Administration (FDA) has ruled that rBST is harmless to people, although critics point out increased levels of insulin-like growth factor 1 (IGF-1) in milk produced using this hormone. The use of rBST is banned in Canada, parts of the European Union, Australia and New Zealand.

NUTRITION

Nutrition plays an important role in keeping cattle healthy and strong. Implementing an adequate nutrition programme can also improve milk production and reproductive performance.

Nutrient requirements may not be the same depending on the animal's age and stage of production. Forages, which refer especially to hay or straw, are the most common type of feed used. Cereal grains, as the main contributors of starch to diets, are important in meeting the energy needs of dairy cattle. Barley is one example of grain that is extensively used around the world. Barley is grown in temperate to sub-artic climates, and it transported to those areas lacking the necessary amounts of grain. Although variations may occur, in general, barley is an excellent source of balanced amounts of protein, energy, and fibre.

Ensuring adequate body fat reserves is essential for cattle to produce milk and also to keep reproductive efficiency. However, if cattle get excessively fat or too thin, they run the risk of developing metabolic problems. Scientists have found that a variety of fat supplements can benefit conception rates of lactating dairy cows. Some of these different fats include oleic acids, found in canola oil, animal tallow, and yellow grease; palmitic acid found in granular fats and dry fats; and linolenic acids which are found in cottonseed, safflower, sunflower, and soybean. It is also important to note that proper levels of fat also improve cattle longevity.

Using by-products is one way of reducing the normally high feed costs. However, lack of knowledge of their nutritional and economic value limits their use. Although the reduction of costs may be significant, they have to be used carefully because animal may have negative reactions to radical changes in feeds. Such a change must then be made slowly and with the proper follow up.

PESTICIDE USE

A survey of the primary dairy producing areas in the US indicated that 13 per cent of lactating animals were treated with insecticides permethrin, pyrethrin, coumaphos, and dichlorvos primarily by daily or every-other-day coat sprays. Workers, particularly in stanchion barns, may be exposed to higher than recommended amounts of these pesticides.

BREEDS

In the United States, dairy cattle are divided into six major breeds. These are the: Holstein-Friesian, Brown Swiss, Guernsey, Ayrshire, Jersey, and Milking Shorthorn. In Rajasthan, an indigenous breed called Tharparkar exists, named from the Tharparkar District, now in Sindh Pakistan. Another type of dairy cow known as Nagauri from Nagaur District, the bull of which is renowned for its ability to plow fields and run. Traditionally, they used to pull covered wagons, known as rath, and in marriages to transport the newlywed couple. They are now a crutch for thriving agricultural and livestock rearing societies of the Thar Desert. Many other breeds are used nearly exclusively for beef, or for both dairy and beef purposes.

3

Livestock Farming: Brief Description of the Sector

The use of cattle, sheep, goats, pigs, poultry, and other livestock offers many benefits to millions of farmers in the developing world. These animals are integral to rural livelihoods and culture, providing food (meat, blood, eggs and dairy products), materials (wool, hide, horns, etc.), income, and mechanical power for pulling carts or plowing fields. Asia has been identified as the developing region in which the demand for livestock products is expected to rise most rapidly.

Livestock manure can serve as a source of fertilizer. Grazing can help sustain vegetation and promote biodiversity by dispersing seeds, controlling shrub growth, breaking soil crusts, stimulating grass growth and improving seed germination. Livestock also may represent savings and currency, and have cultural value as well. For example, gifts of livestock may serve to resolve conflicts or cement marriages.

Livestock production can be categorized under three main systems: grazing, mixed farming and industrial:

1. Grazing systems generally rely on native grassland or forests for fodder, with little or no use of crops or imported inputs, and are traditionally managed by pastoralist communities.
2. Mixed farming systems integrate livestock and crop production. Adding livestock to their farms helps farmers diversify risk and extract value from otherwise valueless or low-value by-products of each activity: crop residue becomes feed, manure becomes fertilizer. Soil nutrients can be further replenished by rotating leguminous (nitrogenfixing) fodder crops with food crops. These systems are managed by settled farmers.
3. Industrial production systems concentrate livestock populations in special facilities and separate their feeding and waste processing from the land on which they live. Feed is provided directly instead of being

acquired through grazing, and manure is transported off-site. Generally, these systems are owned by relatively wealthy individuals and managed by local employees.

Grazing systems are most favoured in arid, semi-arid, or other areas of marginal value for crop-based agricultural production, while mixed farming systems flourish in temperate, subhumid, humid, and some highland climates. Industrial production, because it does not depend on local fodder supplies, can be conducted in any climate and generally occurs near the urban centers it supplies, sometimes even in peri-urban areas.

Although more slowly in sub-Saharan Africa than in most other regions. This increase is driven by growing population, increasing urbanization and rising incomes. This situation is expected to continue throughout the next decade. A shift towards industrial production—farming of monogastric species (pigs, poultry) fed with grain—may be an unavoidable trend in areas with rapidly growing demand for animal food products. Properly managed, livestock production can enhance land and water quality, biodiversity, and social and economic well-being.

However, when improperly managed, livestock production may cause significant economic, social and environmental damage. Increasing livestock production has the potential to increase environmental harm. This guideline is intended to help identify potential adverse environmental impacts and to suggest mitigation and monitoring options, as well as "best management practices" to address them.

POTENTIAL ENVIRONMENTAL IMPACTS IN THE LIVESTOCK SECTOR AND THEIR CAUSES

LARGE AREAS OF LAND DEGRADED

Overgrazing

Overgrazing of rangeland reduces the density of vegetation and the amount of organic matter generated. This, in turn, increases soil erosion from wind and water and decreases soil fertility through loss of nutrients. In arid and semi-arid areas these impacts may also contribute to desertification. Fortunately, ecosystems in these areas demonstrate considerable resilience and often recover when grazing pressure is reduced, either through traditional methods or through modern management practices.

Policy and Legal Problems

In areas that traditionally rely on grazing, the health of rangeland is generally best maintained by traditional pastoralist practices, which regulate grazing location and herd size in accordance with drought cycles and the supply of fodder. In fact, government policies or donor interventions that disrupt or discourage these practices may be a root cause of degradation. A variety of government policies may restrict the movement of livestock within a range area and prevent livestock managers from moving stock from areas that have been depleted of fodder to better supplied areas.

Two particular policy-based problems are:

1. *Land Tenure Insecurity:* Lack of confidence in secure title to rangeland (especially on communal lands) has been shown to reduce the incentive to manage the land sustainably. Many national governments have either implicitly or explicitly claimed ownership of range and wildlands and ignored traditional or customary claims.
2. *Privatization of Communal Resources:* Where national governments have privatized, or are privatizing, formerly state-owned or communal lands, new owners may erect fencing or prevent herds from crossing or grazing on their property.

Wells and Boreholes

Traditionally, access to water on critical grazing lands has been controlled to limit livestock populations and prevent herds from outgrowing the forage supply in dry areas. Thus, new wells or boreholes, generally sponsored by donors or governments, may undermine traditional livestock management systems practices by allowing herds to grow beyond sustainable levels for surrounding areas.

Overgrazing and degradation are most noticeable in the immediate vicinity of the boreholes or wells, but their effects can extend (in gradually decreasing severity) over a considerable radius. Boreholes also reduce pressure on livestock owners to decrease herd size during drought and may discourage movement of herds to other rangelands, disrupting historic wet season/dry season grazing patterns. Larger herd sizes and reductions in pastoral movement may prove to be a recipe for severe degradation of soil and vegetation.

Wet-season Grazing

Poor timing in the use of rangeland can also damage the soil. Wet-season grazing can compact the moist earth, reducing its ability to absorb moisture This increases erosion from water run-off.

Poor Balance of Livestock Species

Each species or breed of livestock has foraging preferences and will graze favoured areas and plants while neglecting others. Browsing animals, such as goats and camels, prefer the leafy tops of shrubs. By contrast, grazers tend to consume ground-level grasses and leafy plants. A poor balance between browsers and grazers can change the mix of plants in ways that degrade the area. For example, too many grazers can diminish the number and populations of herbaceous plant species and allow woody plants to become dominant, changing, possibly irreversibly, the character and utility of the ecosystem.

DAMAGED HABITAT AND REDUCED BIODIVERSITY

Livestock production can damage habitats and reduce biodiversity in wildlife and domestic stock, in vegetation, and in aquatic and wetland ecosystems.

Harm to Wildlife and Domestic Stock and Loss of Wildlife Habitat

The loss of habitat caused by livestock production in grazing and mixed farming systems may be one of the greatest threats to wildlife. Human population growth and density, and the accompanying increase in livestock, often leads producers to expand livestock grazing ranges into wild lands and the conversion of wild lands to mixed farming use.

Slaughter of Wildlife by Livestock Managers

Another danger to wildlife is intentional slaughter by livestock managers. Fear that the wildlife will prey on livestock and damage crops is a common motivation, as is the belief that the wildlife are competing with livestock for fodder, the desire to prevent spread of disease to livestock, and concern for human safety.

Extinction of Local Livestock Breeds

Systematic livestock production may result in loss of genetic diversity in livestock species. This is unfortunate because genetic diversity is a measure of a species' robustness. Local breeds may have traits conferring resistance to emergent or future pathogens, or have other favourable adaptations to local environments. The consistent replacement of local breeds with more productive imported ones can contribute to the extinction of that breed and of all the genetic diversity harboured within its population.

HARM TO VEGETATION

Clearing of Forest and Wild Lands

Vegetation is typically altered or destroyed when forests/wild lands are cleared or are burned to promote new growth. This changes local ecosystems and may contribute to global warming. Fires to burn vegetation are dangerous and degrade air quality.

Loss of Rangeland Fertility

Ironically, mixed farming systems may reduce the fertility of rangeland while helping to solve a farmland problem. Traditional farming practices cause a net loss of nutrients in farm soils; that is, when crops are harvested and sold (or even when human waste is deposited in a latrine) nutrients that make the soil fertile may be lost.

Mixed farming reduces the extent of this loss, by transferring nutrients from the range to the farm in the form of manure. The gain in fertility for the farm is, of course, a net loss for rangeland. Over time, the altered nutrient balance can reduce the productive capacity of the range and/or lead to changes in the composition and density of plant species.

Loss of Farm Fertility

Damage from mixed farming often occurs when poverty and population growth pressure change the crop/grazing land ratio, where other nutrient sources are not available. As the land area available for each grazing animal shrinks, overgrazing becomes prevalent leading to soil erosion and nutrient loss. Harvesting livestock further reduces nutrients available for crop production.

Damage to Riparian Soil and Vegetation

Livestock in grazing and mixed farming systems often graze very heavily in riparian areas along streams and lakes. Results include trampling, loss of vegetation, soil disturbance, and soil compaction, erosion and/or sedimentation which can severely damage riparian habitats, increase siltation and adversely affect watersheds. Pesticide contamination from treatments to protect livestock from insectborne infections (*e.g.*, livestock "dipping") may ultimately reach the aquatic environment. Here it can be toxic to aquatic organisms, as well as people or animals who depend on these sources for drinking water.

Introduction of Invasive Plants

New breeds or fodder crops can introduce invasive non-native plants into a region. The manure, coats and hooves of newly introduced breeds can carry plant seeds. Most non-native plants are not invasive, but when they are, the results can be devastating.

Contamination from Manure

Livestock manure contains relatively high concentrations of nutrients, solids, enteric bacteria and other microorganisms, and organic material. The manure from industrial livestock operations is often discharged or "leaked" into lakes or streams, because it cannot be economically transported to replenish crop fields. When this occurs, the nutrients can cause eutrophication (rapid plant growth in water bodies), solids can create sedimentation, and organic material leads to oxygen depletion (BOD) of the water. Manure from mixed farming, if applied in a concentrated fashion, can lead to similar problems.

Degrade Water Quality and Reduce Water Supplies

Where water is scarce, either chronically or seasonally, the diversion of water to sustain livestock potentially limits its availability for other purposes. This is of particular concern in arid and semi-arid regions, where the construction of boreholes to supply livestock can lead to unsustainable withdrawal rates and the dangerous depletion of aquifer reserves.

Stockpiled manure can contaminate bodies of water, causing myriad adverse effects. These include eutrophication, oxygen depletion, sedimentation, contamination with enteric bacteria and possibly other pathogenic organisms, toxic pollution from pesticides, and contamination of groundwater and aquifers with both nitrates and pesticides. Moreover, high concentrations of nitrate in potable water supplies represent a potential health hazard, especially for children.

Harm to Human Health

Using water for farm animals may make it less available for the many uses that influence health, such as bathing, washing, cooking, and drinking. Moreover, excessive contamination by enteric microorganisms, toxic pesticides or nitrates in may render water unfit for human consumption and may be especially dangerous to children.

Pesticides or other vector control treatments used on livestock represent threats to the health of livestock managers, their families, and others exposed directly or through water use. These substances may be toxic, cause birth defects, alter children's proper development, promote cancer, or slowly poison one or more organ systems.

Climate Change/Global Warming

Approximately 17 per cent of global methane is produced by livestock digestion and manure. Methane is a greenhouse gas that has 56 times more global warming potential than carbon dioxide over a 20-year time horizon. Furthermore, clearing wild or forest land for new fields reduces the land's ability to act as a carbon sink. Setting fire to vegetation to clear land or promote new growth for livestock to graze on may also contribute to global warming.

Odour

Concentrated manure stored at industrial livestock facilities can generate strong and unpleasant odours, damaging the quality of life of nearby residents. This problem is most evident when facilities are located in densely populated areas.

SECTOR PROGRAMME DESIGN—SOME SPECIFIC GUIDANCE

The following questions and suggestions are intended to help project designers and managers identify factors and practices that may cause—or prevent—adverse environmental impacts. Bear in mind that the first priority of most livestock managers and farmers is household food security and family welfare. Sustainable practices must always be balanced against these immediate demands.

CONSIDER CLIMATE, TERRAIN, AND ECOSYSTEM

Since environmental impacts from livestock production vary, depending on the specific climates, terrains and ecosystems involved, project designers need to address these characteristics during the initial design phase:

- What is the climate in the project area (arid, semi-arid, temperate, subhumid, humid)? What do historical records of rainfall patterns and flooding indicate? Is the proposed livestock management practice compatible with this climate?
- What terrains are found in the project area(s) (alluvial plain, highland, rocky desert, wetland, etc.)? Do they have any known vulnerabilities to livestock grazing? For example, are there many unprotected streams or rivers? Are there slopes with limited topsoil sensitive to erosion?
- Will the project encompass or border on protected or ecologically sensitive areas? Are there any threatened or endangered species in the area? Would the proposed project directly or indirectly threaten wildlife or native vegetation? For example, does the project require expansion of grazing into protected areas or make livestock more vulnerable to wildlife predators, triggering reprisals by farmers?

EVALUATE POLICY, LEGAL, CUSTOMARY AND CULTURAL CONTEXT

The policy, legal, and cultural contexts of a project merit attention, since, these factors may limit programme options or erect substantial barriers to success.

Policy/Legal

- Do livestock owners and managers have legal and recognized ownership and responsibility for land and grazing resources? What are the current or proposed national land tenure policies, and how are they implemented at the local level? Are they effective in encouraging sustainable management of grazing land and resources?

- What is the tenure status of current or proposed rangeland—Is it owned by individuals or the community? Does the government have claims?
- What wildlife protection laws exist and how are they implemented in the region?

Culture/Custom

- What role does livestock play in local culture and customs, and how might the proposed project affect these practices? Would the proposed project disrupt traditional grazing patterns?
- If there are customary land tenure arrangements, what are they and how would the proposed livestock management system work within these arrangements? For example, will livestock herders—who often come into conflict with farmers, particularly during droughts—have a means of working out disputes with farmers?
- If livestock management arrangements are communal, how would these be affected by and/or affect the proposed development activities?

ASSESS CURRENT AND PROPOSED SPECIES AND BREEDS

Introduction of a new breed into an area should be approached with caution. The new breed may bring with it diseases that can decimate local livestock herds and wildlife. In addition, the foraging habits of a new breed may lead to a sharp decline in available forage and biodiversity. A new breed's reproductive habits can lead to a herd's uncontrolled growth.

Weeds can be accidentally introduced along with a new animal species, and they may displace desirable vegetation. The long term full costs and benefits of introducing a given new livestock species into a particular environment should be assessed.

For example, large animals roam over extensive areas in search of food, often require a greater financial investment, can be more difficult to control, and have lower reproductive potential than small animals. It is important not to underestimate the value of breeds that are well adapted to the environment.

Livestock tend to overgraze favoured areas and plants while neglecting others. Native plants may not be able to survive heavy grazing. While unforaged plants tend to lose vigour and nutritional value as they mature. The introduction of new plant species (whether accidentally or intentionally) may quickly result in replacement of native plants. Even when grazing pressure is reduced, exotic plant species sometimes retain their dominance. Ask the following questions when introduction of a new breed or species is proposed:

About Current Species and Breeds

- Which wild and domestic species are already present in the area, and in what numbers?
- How have they been used in local farming systems and traditions?
- What are the feeding preferences of local livestock and wildlife? What is the balance between browsers and grazers? Do domestic species compete for resources with one another and with wildlife?
- Have population sizes of wild or domestic species changed recently?
- Could local breeds satisfy the project's needs?

About Proposed Species and Breeds

- If new species or breeds are being considered, how will their production complement or conflict with local species or breeds, wildlife, and other local resource users?
- How would they fit in local herding systems?
- Are they well suited to the local climate and environment?
- Are they resistant to local livestock diseases?
- Have alternative species or breeds been considered for possible introduction?

EVALUATE CURRENT AND PROPOSED LIVESTOCK MANAGEMENT PRACTICES

To maximize forage productivity, it is best to combine or alternate various livestock breeds on a range. Their differing food preferences can help to keep plants productive by minimizing overgrazing of a particular favoured area and allowing less preferred plant species time to mature. It is prudent to make superior forage available to those animals with the highest needs. When forage is limited, livestock managers may decide that young and milk-producing animals must have first access to new pastures and ranges with a wide variety of abundant forage.

Managers should investigate the value of various systems of rotating livestock. Rotation allows land to be grazed continuously throughout the year. Livestock can be rotated between fields or ranges to prevent the buildup of disease and to vary grazing pressures. Through either fencing or herding, they can be relocated into croplands to consume crop residues. Assessment of seasonal grazing patterns should include potential impact on soils.

Dry-season grazing can benefit the land by breaking up crusted soil and working seeds into the ground. By contrast, grazing on moist soil can cause considerable soil compaction, which reduces the soils ability to absorb moisture and can result in increased erosion from run-off during the rainy season. Many of the environmental impacts from livestock production are associated with particular practices of livestock managers.

Thus it is critical to understand current practices and how the proposed project might alter these practices or promote new ones:

- Who are the local community's livestock managers?
- What practices do a family or community use to control the size and composition of livestock herds?
- How do livestock managers currently control livestock movement? Will the proposed project change these movements in a way that might harm the environment?
- Does the proposed project require the construction of fences? If so, will they interfere with wildlife migration or transit of livestock belonging to other communities? Could the fences lead to overgrazing and land degradation? Will the fences be built with local materials? Would living fencing be practical? Would solar-powered electric fencing be technically and economically feasible?
- Are streams and riverbanks currently protected from livestock damage? If the proposed project will open new areas to grazing, will water supplies need to be protected?
- Must steps be taken to prevent new livestock and associated animals (*e.g.*, dogs) from transmitting disease to wildlife? Is there a vaccination/animal disease control programme available for this purpose?
- Will the project involve construction of improvements (*e.g.*, boreholes or other infrastructure)? Could these lead to unplanned changes in herding patterns and overgrazing?

ASSESS DEMAND AND USE OF LIVESTOCK PRODUCTS

- Who is marketing livestock and livestock products?
- Is the demand for livestock products coming from local or outside populations? How rapidly is it increasing or decreasing? How stable is the demand?
- In preparing livestock products, are people using technologies which reduce impacts on the environment, open additional markets, or improve health and nutrition (of people and animals.)

ASSESS LIVESTOCK ECTOPARASITE MANAGEMENT

Epidemic and endemic diseases continue to be a major constraint to livestock productivity in large parts of the developing world. Although vaccines have controlled many of the epidemic diseases, they continue to cause severe economic losses through morbidity and mortality. These diseases include the infections caused by vector-borne haemoparasites and helminths. Existing technologies, such as chemotherapeutic agents and live vaccines that were previously successful in controlling these diseases, are no longer effective—because of acquired resistance or weakened delivery services. Appropriately designed alternatives are often lacking.

CONSIDER POPULATION PRESSURE AND DISEASE BURDEN

Two factors that may affect the outcome and impact of projects population growth and fatal or debilitating epidemic diseases. Population growth may lead to conflicting uses of grazing lands or reduce individual farms or rangeland to sizes that cannot sustain livestock. Fatal or debilitating epidemic diseases may weaken effective dissemination or replication of proper livestock management techniques.

- What is the current and projected population growth rate in the project area? How might this affect project sustainability in the future?
- What is the current extent of the HIV/AIDS epidemic in the region? How might this affect the composition of the population (size, ethnic makeup, age/gender distribution) and family structures necessary for project sustainability? How will development and livestock technical support services be affected?
- Are there other epidemic diseases in the region, such as sleeping sickness, that might adversely affect project implementation?

LIVESTOCK MANAGEMENT

LIVESTOCK PRODUCTION

Areca Sheath

Areca (*Areca catechu*) is a commercial crop and the fallen areca sheath is a good source of dry roughage to animals. The nutritional value of areca sheath is similar to paddy straw (CP, NDF, ADF) with an advantage of having less lignin, silica and more calcium, sulphur and copper. Feeding of areca sheath to sheep and dairy animals replacing paddy straw could be an alternate approach.

Lignin Degradation

Activities of lignolytic enzymes produced by immobilization on areca sheath recorded temperature and pH stabilities over a wide range, thus making it desirable for supplementation in animals.

Feed Additives

Out of the 44 plant products, *viz.* seed pulp (5), leaves (20), spices (13), citrus fruit peel (3), oils (3), and 23 plant combinations, screened for their anti-methanogenic activity using *in vitro* gas production test, a mixture of plant products and yet another combination of four plants appeared to have good potential as feed additives in the ruminants. Use of nitrate reducing bacteria as probiotic exhibited 80 to 99 per cent reduction of *in vitro* methane emission.

There was no accumulation of hydrogen, nitrate or nitrite in the incubation medium, and no adverse effect on *in vitro* true digestibility of feed. Supplementation of Chicory root powder, *Lactobacillus acidophilus* as symbiotic in the diet of dogs proved beneficial in terms of fibre utilization, calcium and lipid metabolism, hind gut fermentation and cellmediated and humoural immunity. *Helianthus tuberoses* L.*Jerusalem artichoke Hathi choke*; a novel phytogenic-origin probiotic, showed immense potential for use in the diet of dogs for enhancing their digestive health, lipid profile and immunity.

Unconventional Feed

Detoxification procedures for jatropha, *karanj* and neem seed cakes were standardized in an industrial process. In growing lambs, replacement of soybean meal with detoxified jatropha cake up to 37.5 per cent showed no adverse effect on DM intake, nutrient utilization, bloodbiochemical profile, growth rate and feed conversion efficiency.

Nutritive value of *Robinia pseudoacacia* in terms of high organic matter, crude protein and low lignin content was comparable to *Grewia optiva*, and showed higher nutritional values than other tanniniferous forages. Combined treatment with curry and bel leaf powder had positive effect in restoration of cyclicity and fertility in acyclic goats and buffaloes.

Chelated Minerals

Cell-mediated immune response was significantly higher in kids given organic zinc (40 ppm zinc as zinc methionine) as compared to zinc sulphate.

Nutritional Manipulation

Feeding growing Frieswal bull calves with three different levels of rumen degradable protein and rumen undegradable protein for achieving a targeted body weight gain of 700 g and 1,000 g/day, yielded only 608 g and 772 g of average daily gain suggesting that there is a need to develop our own feeding standards for higher body weight gains.

Complete Pellets

Loin eye area increased linearly with increased level of concentrate mixture in different supplementation group. Kids fed diet without any concentrate mixture had lower yield of separated lean, fat and bone content. Different levels of supplementation improved meat bone ratio of Barbari kids under intensive system.

YAK

Two complete feed blocks (CFB) were prepared based on tree leaves (50 per cent tree leaves, 43 per cent concentrate and 7 per cent molasses) and paddy straw (50 per cent paddy straw, 43 per cent concentrate and 7 per cent molasses). The result indicated that tree leaves based CFB supplemented with limiting micronutrients can be fed during winter to overcome the weight loss in yaks.

POULTRY

Feeding Schedules for Poultry

In PB-1 females, ME restriction was successful in achieving the desired body weight (83 g/week and 5. 75 per cent growth rate) at 20 weeks of age. *Karanj* cake in the diet adversely affected Krishibro chicks, but feeding of detoxified cake improved the performance. The production performance of PB-2 line was better with organic Zn supplementation at 60 ppm and hatchability at 80 ppm. Trace mineral retention in tissues of broiler chicks was linearly related to their level in diet. Herbal antifungal compounds, *viz. Piper longum*, *Ferula assafoetida* and *Cymbopogon citratus* oil were identified as potent antifungal herbal compounds against *Aspergillus parasiticus* in poultry feed under *in vitro* conditions.

Quality Protein Maize in Diet

Dietary replacement (50 to 100 per cent w/w) of normal maize (higher in lysin and tryptophan by 44 and 33 per cent, respectively, over normal maize) resulted in 6.93 per cent higher body weight gain, 7.69 per cent improvement in feed utilization efficiency and 5.15 per cent higher breast meat yield in broiler chicken.

Amelioration of Heat Stress in Broilers

Amelioration of heat stress in broiler chickens was achieved through supplementation of vitamin C and potassium chloride in diet. Heat shock protein 70 (HSP70) was detected in muscle and blood lymphocyte in birds as potential marker of heat stress.

REPRODUCTION

Cattle

Good quality fresh and post-thaw semen of crossbred bulls had significantly higher percentage of HOST positive spermatozoa with low LDH and GOT enzyme leakage as compared to poor quality semen. Frozen semen doses (166,610) of Frieswal bulls were produced and 73,205 doses were distributed that included 50,725 to military farms, 9,000 to field progeny testing project, and 13,480 to various developmental agencies and farmers for cattle improvement.

Buffalo

Buffalo sperm membrane is highly prone for lipid peroxidation as compared to cattle, and in a study on nutrition and male fertility, it was observed that PUFA feeding may reduce lipid peroxidation and improve frozen semen fertility. Studies on nutrition-endocrine interactions in attainment of puberty in female buffaloes showed that supplementation of higher dietary energy (through bypass fat) has a positive effect on early maturity of hypothalamo-pituitary—ovarian axis in terms of early and better LH secretion and ovarian follicular activity in female buffalo calves.

Studies on physiological reasons for low reproductive efficiency and early embryonic mortality in buffaloes showed that antioxidant enzymes, Cu and Zn-SOD are induced significantly during early pregnancy and may be important for overcoming oxidative stress that otherwise would lead to early embryonic mortality. LPS and TNFα appear to modulate prostaglandin production in endometrium and corpus luteum and are important for protecting and enhancing function of corpus luteum to prevent early embryonic mortality.

IGF-I also stimulated progesterone secretion from the corpus luteum. Five proteins, specific to uterine secretions of pregnant buffalo were recognized and it may be possible to use one of these for early pregnancy diagnosis to monitor pregnancy and prevent early embryonic mortality.

Goat

Study on effect of dextran, a new cryoprotectant, revealed that dextran alone did not prove to be an efficient cryoprotectant, but 3 per cent glycerol dextran in combination with 6 per cent glycerol resulted in 39.33±1.45 per cent post-thaw recovery. Hormone delivery system for synchronization of heat in goats was standardized by using sponges of different sizes and shapes. The estrus response, onset of estrus and duration of estrus following use of intra vaginal sponge pessaries for induction of multiple birth during peak breeding activity were 100 per cent, 29.14±4.42 h and 25.71±0.37 h, respectively, with 57.14 per cent conception rate.

The kidding rate following induction of multiple birth was 2.0. During low breeding activity season (summer) the estrus response, onset of estrus and duration of estrus were 71.42 per cent, 60.0±16.54 h and 26.4±11.63 h, respectively, following use of intra vaginal pessaries. Oocytes were recovered by follicle puncture technique for *in vitro* maturation, fertilization and culture from goat ovaries. The average recovery of oocytes using follicle puncture was 1.76/ovary.

The oocytes were cultured in tissue culture medium (TCM-199) supplemented with EGF (10 ng/ml) and fertilized *in vitro* with a cleavage rate of 41.55 per cent. Embryos were bisected through microtools. Out of 25 bisected cells 3 single cells divided into double cells whereas one bunch of 3 cells developed into 4 cells. Adaptability study on goats revealed that water intake of goats kept under asbestos sheet roof sheds was significantly higher compared to the thatch roof shed during both the moderate and cool seasons. Milk production of animals kept inside the thatch roof shed was significantly higher than the animals inside the asbestos roof shed.

Camel

Physiological parameters (rectal temperature, respiration rate, pulse rate) were higher during evening as compared to morning. The morning and evening variation in the THI value was significant for all weeks in different months from October to March. The Benezara coefficient of adaptability (BCA) of camel revealed that BCA was significantly higher during evening as compared to morning. Non-pregnant female camels were divided in groups A,B,C.

Each group was tested with poll gland secretion collected in breeding season (PGS-BS), poll gland secretion collected during non-breeding season (PGS-NBS) and water. The results indicated that the PGS-BS evoked greater olfactory, contact and flehmen response than PGS-NBS and water. Nine female camels, which did not conceive during breeding season were given injection of highly purified hCG for infertility treatment at the time of mating. Pregnancy was diagnosed in three out of nine camels.

Yak

On application of the vaginal sponge containing 600 mg progesterone for synchronization of oestrus followed by AI in yaks resulted in 100 per cent oestrus response with 60 per cent conception rate. The yaks do not experience cold stress (environmental stress) during winter but they are subjected to heat stress (during summer) when the ambient temperature exceeds 10°C at the altitude of 2,750 m above mean sea level. The body weight gain was significantly higher during summer than winter in adult yaks.

The lactating yak cows lost about 5.84 per cent of their body weight during winter along with reduced milk yield mainly due to shortage of feed and fodder. Estrous was displayed year round but the highest incidence (around 65 per cent) was restricted to warmer part of the year. Physiological responses, *viz.* rectal temperature, respiration and pulse rates of yak calves, bulls and lactating cows were recorded during winter and summer. The physiological responses of calves were significantly higher than bulls and lactating cows during summer. Standardization for heat shock protein 70 (Hsp 70) assay was carried out in yaks.

Mithun

Different protocols for preservation of mithun semen indicated that it is possible to preserve mithun spermatozoa at refrigeration temperature in tris-egg yolk diluents till 108 h, and doubling the concentrations of fructose in the diluent buffer increased the progressive motility, live sperm count and decreased total sperm abnormalities. Hence, addition of extra-energy source (fructose) might increase the storage time of mithun sperm at refrigerated temperature up to 5 days.

Pig

Gloved hand method of semen collection was standardized. Semen preservation was done for Hampshire, Duroc and Ghungroo pigs for first time in Asom. The AI technology was used in the farm and in the nearby villages and highest litter size (15 piglets) at birth was recorded from a Ghungroo sow in the farm.

Poultry

Exposure of birds to 650 nm and 475 nm wavelengths of light improved egg production by 6.55 and 3.05 per cent, respectively, compared to 450 nm (incandescent) light.

LIVESTOCK PROTECTION

Foot-and-mouth Disease

During the period under the report, 799 outbreaks were recorded throughout the country. Maximum outbreaks were recorded in the eastern region where there was increase in the number of outbreaks compared to last year.

Disease incidence in West Bengal directly affects the situation in NE region due to animal migration. Drastic reduction in outbreaks was noticed in southern region. FMD cases were not reported in Tamil Nadu, whereas Himachal and Punjab recorded a single case of FMD each. Maximum incidence of disease was reported in March and from August to November.

Out of 1,624 clinical specimens collected from 799 outbreaks, the virus could be detected in 1,067 samples (serotype O-991, A-38 and Asia 1-38) by using sandwich ELISA and multiplex PCR. Serotype O was responsible for maximum number of outbreaks (568) followed by A (33) and Asia 1 (9). Increase in serotype O incidence was noticed in all regions except northern and western region. Serotype A, which was absent in northern and central region last year appeared this year, making it the second most prevalent serotype affecting geographically all the regions of India.

Serotype Asia 1 was completely absent in southern, central and north eastern regions. Incidence of Asia1 serotype increased in western region. Though majority of the outbreaks involved cattle, disease was also reported in buffaloes, pigs, sheep and goats. A major shift was observed in genetic lineage of serotype O viruses circulating in India. The 'Ind2001' lineage viruses gained upper hand after a gap of 8 years and outcompeted PanAsia II lineage.

The cases due to PanAsia II, PanAsia 1 and Branch C-II viruses were not completely absent during the year. Branch-C II lineage viruses responsible for outbreaks during 2008 in West Bengal were recorded in Tripura, Asom, West Bengal and Odisha during 2009–10. In serotype A, it was observed that both VP359 deletion and non- deletion mutants belonging to genotype VII are cocirculating in the field in recent times.

Precisely, the viruses from Punjab and Uttarakhand shared ancestry and clustered in the same VP359 deletion lineage, VIIf, whereas viruses from two different outbreaks from Odisha and those from Andhra Pradesh clustered together and revealed no deletion in the VP3 coding region. During the period, outbreak due to Asia1 serotype was recorded in Gujarat, Madhya Pradesh, Maharashtra, Uttar Pradesh and West Bengal.

The Asia 1 field isolates were grouped with lineage C reiterating the supremacy of this lineage in the field since 2005. In vaccine matching exercise 33 virus isolates were subjected to one-way antigenic relationship study. All the isolates showed close antigenic match with respective in-use vaccine strains indicating appropriate antigenic coverage. Virus isolates (133) including 97 type O, 12 type Asia 1 and 24 type A field isolates were added.

In National FMD Virus Repository 1,687 (1,096-O, 317-Asia 1,259-A, 15-C) well characterized field isolates are available. Under National FMD sero-surveillance testing of 11,560 serum samples from 13 states revealed the presence of protective antibody titre in 42.3 per cent, 40 per cent and 30.5 per cent animal population against serotypes O, A and Asia1 respectively. This baseline response has improved due to application of vaccine. The current in-use vaccine strain IND R2/75 is found still the best and covers all the type O circulating outbreak strains in the country.

Classical Swine Fever

Classical swine fever (CSF), the topmost diseases in pigs, is one of the OIE listed diseases. The CSF or hog cholera is associated with high mortality, abortion, mummification of fetuses. It is of great concern in areas, where the pig population is in high density.

CSF virus isolates maintained at the repository of PD-ADMAS, Bengaluru, were revived and processed for RT-PCR of NS5B genomic region. Phylogenetic tree based on the alignment of 408 nts of NS5B region was constructed. It showed three major branches, each representing three previously described genetic groups with further sub-branches.

The majority of the Indian CSFV isolates belong to subgroup 1.1. Clustering of the Indian field isolates away from the vaccine strain and other related strains revealed that the field isolates did not have any vaccine origin. Phylogenetic analysis based on 52 UTR classified subgroup 1.1 Indian isolates into two minor subbranches, however, no such classification could be seen from the analysis based on NS5b region; nevertheless in the latter analysis grouping of isolates was much more resolved. Lone isolate, IND294/08, is clustered within subgroup 2.2, ak into its grouping based on 52 UTR sequence.

Brucellosis

Serum samples from goat and sheep received from Gujarat, Rajasthan, Uttar Pradesh and Maharashtra revealed that among them 6.35 per cent were positive from brucellosis. Similarly, 1.8 per cent sheep and goat sera samples for Karnataka were positive for brucellosis. Porcine brucellosis is one of the most important emerging zoonoses and currently, no control programme for porcine brucellosis exists and the epidemiological situation is still unknown.

Indirect-ELISA for screening brucellosis in swine was standardized using smooth lipopolysaccahride antigen from a standard strain *Brucella abortus* S99; 22.6 per cent were found positive, which is much higher than that found in sheep and goat. An indirect ELISA was developed to detect *Brucella* in any of the susceptible species. Expression of four recombinant *Brucella* proteins, *viz.* BP26, BP39, superoxide dismutase (SOD) and *Brucella lumazine* synthase (BLS) is being envisaged to overcome the problem of cross reactivity in diagnosis of this disease.

Leptospirosis

PCR based techniques were developed for differentiation of pathogenic and non-pathogenic *Leptospira* as well as for typing. Initially, *Leptospira* genus specific PCR was carried out to differentiate the *Leptospira* from other spirochete organisms, *viz. Borrelia*, *Treponema* and *Leptonema* etc., which amplified 331 bp products from the *Leptospira* organism only.

Sero Prevalence of IBR

Serum samples collected from Andhra Pradesh, Gujarat, Jammu and Kashmir, Karnataka, Madhya Pradesh Maharashtra, Meghalaya, Manipur, Orissa, Punjab, Rajasthan, and Tamil Nadu, were subjected to Avidin Biotin ELISA. An overall apparent percentage positive was 34. The highest prevalence of 67 per cent was in Tamil Nadu, and the lowest in Meghalaya. Region-wise positive percentage were as follows: East zone 17 per cent, south zone 24 per cent, west zone 48 per cent, north zone 37 per cent, north-east 39 per cent, and central zone 25 per cent.

Herbal Anticoccidial and Antidiarrhoel Drugs

The extracts CIRG-5 and CIRG-6 from two different plants were evaluated *in vitro* against sporulation of coccidian oocysts and found 96 to 100 per cent effective to check the sporulation up to 96 h. Eight potential plant formulations of herbal extracts were subjected for clinical trials to control diarrhoea caused by *Escherichia coli* in kids of less than 1 month of age. Final product was developed from plant extract prototype CIRG-I and CIRG-2A and final products (formulation) @ of 10 mg/kg bwt for 1–2 days orally was highly efficacious to control EPEC diarrhoea in kid.

VACCINES AND THERAPEUTICS

- Live attenuated orf, buffalopox and camelpox vaccines met the thermo-stability and potency requirements of the biologicals.
- Bi-valent DNA vaccine containing HN and F genes of Newcastle disease virus was evaluated along with genetic adjuvant comprising IL-2, IL-4, IFNγ, and GMCSF gene constructs.
- An aroA gene knockout mutant of *Pasteurella multocida* serotype B:2 was constructed for the development of HS mutant vaccine.
- Chitosan coated cationic PLG nanoparticles were prepared and when tested for the transfection efficiency with commercial transfecting agent, indicated that these nanoparticles can be safely used for delivery of DNA vaccines.
- *Trypanosoma evansi* beta tubulin protein was expressed in both prokaryotic and eukaryotic expression systems.
- Two new target proteins of *Fasciola gigantica, viz.* leucine aminopeptidase and peroxiredoxin coding genes were cloned and successfully expressed in their native conformation in a prokaryotic expression system.
- Immunomodulators of microbial and herbal origin, *i.e. Mycobacterium phlei, Lactobacillus acidophilus* and *Tinospora cordifolia* were effective under field conditions to improve productive performance and mitigate

the adverse effect of sub-clinical caecal coccidiosis and thereby improvement of body weight gain in broiler birds.

- Pharmacokinetics of gatifloxacin revealed that it has therapeutic potential for use in sheep.
- Medetomidine-butorphanol, thiopental and halothane combination provided better clinicophysiological and haemodynamic stability in general anaesthesia in buffaloes.
- Three TLR genes of yak (TLR 5, 7 and 9) and 2 TLR genes of mithun (TLR 4 and 7) were sequenced completely.

DIAGNOSTICS

- A recombinant lipoprotein 'P48' of *Mycoplasma agalactiae* was evaluated in an ELISA as a rapid immunodiagnostic tool in sheep and goats.
- A genus-specific PCR assay was standardized for rapid detection of *Arcobacter* sp. targeting 16S rRNA for confirmation of the isolates of this emerging pathogen.
- A 36 kDa protein was identified in different serovars of *Salmonella* during biofilm formation.
- Magnetic nanoparticles (MNPs) were synthesized by two different processes and were characterized for structural, morphological, size and magnetic properties. The MNPs were functionalized for immunoassay applications and evaluated for cellular toxicity in *in-vitro* experiments.
- A specific and sensitive diagnostic marker was identified for early detection of pancreatitis in dog.

MOLECULAR CHARACTERIZATION OF PATHOGENS/RECEPTORS

- Indian H5N1 isolates were characterized genetically and grouped with EMA *3/2.2.3 subclade,* which branched from two distinct nodes in the phylogenetic tree.
- The alignment of deduced amino acid sequences showed that all the Indian H5N1 viruses characterized contained the multiple basic amino acid motif at their HA cleavage site, which is characteristic of highly pathogenic avian influenza viruses.
- Two viruses from 2010 outbreaks in West Bengal contained amino acid asparagine at position 31 of the M2 protein, suggesting that both viruses are resistant to drug amantadine.
- An isolate of buffalo pox from man and animal recovered from an outbreak from Pune was characterized for B5R, H3L, D8L and A27L genes.

- An orf isolate from camel in a sporadic infection was characterized for B2L, GIF and VLTF-4 genes.
- Genotyping and partial sequencing of rotaviruses from man and pigs revealed human-pig reassortants as well as presence of novel combinations in India.
- Complete HN gene of velogenic Newcastle disease virus containing kozak sequence was amplified and cloned into pcDNA3. 1 eukaryotic expression for enhanced expression.
- Phylogenetic analysis by AP-PCR, PFGE and ribotyping showed considerable similarity between human and fish isolates of pathogenic and pandemic *Vibrio parahaemolyticus,* which reflected the public health risk potentials of fish.
- Molecular characterization by PCR-SSCP analysis of *Oesophagostomum* spp. revealed two distinct conformers with six nucleotide polymorphisms (SNPs) among male and female parasites.
- Probability of evolutionary changes among prevailing genotypes of *Oesophagostomum* spp. was predicted by analysing secondary structure of RNA.

QUALITY CONTROL AND PRODUCTION OF VETERINARY BIOLOGICALS

- 223,500 doses of RD 'F' strain vaccine; 166,000 RD 'Mukteswar' strain vaccine; 132,500 doses of fowl pox vaccine; 184,195 doses of lapinized swine fever vaccine; 423,800 doses of tissue culture sheep pox vaccine; 160,600 doses of PPR vaccine; 57,925 doses of *Brucella abortus* strain-19 (live) vaccine; 3,640 ml HS adjuvant vaccine; 73,500 doses of tuberculin PPD; 50,000 doses of Johnin PPD; 25,250 doses of mallein PPD; 55,250 ml of *Brucella* agglutination test antigen; 6,900 ml of *Brucella abortus* Bang Ring antigen; 19,860 ml of Rose Bengal Plate Test antigen; 219 ml of *Brucella* positive serum; 9,000 ml of *S.* Pullorum coloured antigen; 45 ml *S.* Pullorum positive serum; 87 ml *Salmonella* 'O' serum and 3,750 ml of *S.* Abortus equi 'H' antigen were produced and quality tested.
- 7.32 million monovalent doses of FMD vaccine were produced at Bengaluru campus.
- 435,100 doses of PPR vaccine were produced; similarly, 33 PPR c-ELISA, 22 PPR s-ELISA and 100 doses of goat pox vaccine were supplied by Mukteswar campus.

Sero-monitoring of Important Equine Diseases

During 2009–10, equine disease sero-survey was conducted in Rajasthan,

Haryana, Punjab, Uttar Pradesh, Uttarakhand, Madhya Pradesh, Chhattisgarh and Sikkim. A total of 176 out of 1,045 equines (16.84 per cent) had antibodies to equine influenza virus while 28 (2.67 per cent) were positive for EHV-1 and 107 (10.43 per cent) were positive for antibodies to Japanese encephalitis virus. Antibodies to *Babesia equi* were detected in 320 (38.09 per cent) of 840 equines, while *T. evansi* antibodies were detected in 49 (5.4 per cent) out of 906 equines tested. One case of equine infectious anemia was detected in Uttarakhand. EIA-positive equine was reported after a gap of 11 years from India. None of the 1,045 equines tested showed antibodies to *Brucella* and *Salmonella* Abortus equi 'H' antigen.

Japanese Encephalitis (JE) Virus Isolation

JE, a mosquito-transmitted disease caused by JE virus, is causing disease in equines, pigs and humans. It is prevalent in eastern and southern Asia. Serum samples from clinically affected horses and pooled mosquito vector samples from the surrounding field were found positive for JEV RNA by RT-PCR. Virus was isolated from one of the horses. This is the first report of isolation of JEV from a horse in India. The phylogenetic analysis of the envelope gene (1500 bp) sequence of the isolated virus, JE/eq/India/H225/2009 indicated that the isolate belonged to genogroup III of JEV, and that the equine isolate clustered together with Vellore group of JE isolates from India.

Camelpox Virus Zoonosis

Camelpox virus (CMLV) causes skin eruptions in addition to severe generalized exanthema mostly in camels and outbreaks occur frequently in camel-rearing north-western region of the country. However, their zoonotic potential has not been proven conclusively. The disease was confirmed on the basis of isolation of the virus from diseased camels, detection of the CMLV antibodies in the serum of diseased camels and human patients by serum neutralization test and amplification of CMLV-specific three full-length envelope protein genes (A27L, H3L and D8L) and one partial gene (C18L) of the isolated viruses from camels and human specimens.

The phylogenetic analysis indicated that CMLV isolates cluster closely with variola virus (VARV) and vaccinia virus (VACV). The CMLV zoonosis indicates the changing pattern of host specificity of poxviruses and certainly points towards a declining cohort immunity against Orthopoxviruses in humans, which could be a public health concern. This is the first confirmed report of camelpox virus zoonosis.

Cysteine Proteinases from Trypanosoma Evansi

Parasite derived cysteine proteinases were proposed as potential targets for immunotherapeutic, chemotherapeutic and diagnostic reagents for parasitic diseases. The proteinases of camel isolate of *T. evansi* were analysed by gelatin substrate SDS-PAGE containing collagen antigen confirming the presence of cysteine proteinases in *T. evansi*. Some of these cysteine proteinases were reactive with experimental/field serum sample from equines with *T. evansi* infection. This indicated the potential of these enzymes in serodiagnosis of this economically important disease of livestock.

FISH

Capture Fisheries

The marine fish landings of India during the year 2009 has, provisionally, been estimated as 3. 16 million tonnes. The pelagic finfishes constituted 52 per cent, demersal fishes 28 per cent, crustaceans 16 per cent and molluscs 4 per cent of the total landings. The sectorwise contributions during the year 2009 were mechanized 74 per cent, motorized 22 per cent and artisanal 4 per cent. The west coast accounted for 56 per cent of the total landings. The northeast region, comprising West Bengal and Odisha contributed 20 per cent to the total production; southeast region consisting of Andhra Pradesh, Tamil Nadu and Puducherry contributed 24 per cent. On the west coast, the northwest region comprising Maharashtra and Gujarat recorded 26 per cent of the total, and the southwest region comprising Kerala, Karnataka and Goa contributed 30 per cent.

Cruzing for Hilsa Fisheries in Hooghly

The investigations on distribution of hilsa in Hooghly from Nabadweep to Kakdweep estuary were done through cruzing. It established that depth and width of the river stretch affect the fishing effort and hilsa catch. The zones with <3 m depth had rare fishing activity, while fishing intensity was higher in deeper water (10 m depth).

Hilsa was evenly distributed over the observed >100 km freshwater stretch. Movement of hilsa in shoals was missing during non-spawning season. Hilsa was available throughout the year with peak in monsoon. The observations may lead to possible occurrence of a resident stock of hilsa in Hooghly in addition to regular seasonal influx of migrant groups from sea to freshwater zone for breeding.

Impact Assessment of Fishing Ban

To assess the impact of fishing ban on reservoir ecosystem, massbalanced models were used on different ecological groups before and after introduction of fishing ban in Wyra reservoir of Andhra Pradesh using Ecopath software. It is the first attempt in India on modeling a tropical productive reservoir, which contributes to fishery, particularly the commercially important prawn fishery.

The results of the assessment indicated:

- Slight increase in system turnover,
- Better organization/increased efficiency,
- Decreased tendency for eutrophication (low sum of all flows into detritus),
- Superior stability of total biomass/TST,
- Higher total primary production/total respiration,
- Increased resilience (higher overhead per cent), and
- Higher maturity and stability (higher primary production/biomass and lower ascendancy per cent).

Formulation of Shrimp Feed

Yard trials revealed that 3 per cent Ca with 2 per cent P supplementation resulted in a significantly higher digestibility of Ca and P as well as weight gain compared to other groups receiving lower levels of Ca and P. Feeds for high salinity, indicated that high growth can be achieved at 40 and 33.5 per cent dietary crude protein levels compared to lower dietary protein levels. Based on the better protein efficiency ratio and apparent protein utilization at high salinity (40 per cent), protein utilization efficiency was better at 33.5 per cent dietary protein level.

Reducing Fish Meal in Shrimp Feed

One of the bottlenecks to the approach of using plant proteins for reducing the level of costly fish meal in shrimp diet, is the high cellulose content of these feeds. *Bacillus macerans,* a potential cellulolytic bacteria, isolated from gut of *Lates calcarifer,* was identified for testing as feed supplement in plant protein based shrimp feed.

After 48 h of incubation, crude protein content, free glucose concentration and microbial count were maximum in the feed and dry matter loss was less. Growth, survival, protein efficiency ratio were significantly higher and feed conversion ratio was significantly lower in shrimps fed with feed supplemented with live cellulolytic bacteria as compared to control feed.

FISH FEED FOR MARINE ORNAMENTAL FISHES

The CMFRI has launched a dry formulated feeds named CadalminTM *Varna* for marine ornamental fish like clown and damsel fish. These feeds contain 40 per cent protein, 9 per cent fat, 39 per cent carbohydrates, 7 per cent ash and less than 2 per cent fibre, colour imparting nutrients like carotenoids from natural sources, nutraceuticals and microbial products. They are slow sinking crumbles available in three particle sizes, 0.25 mm, 0.75 mm and 1 mm produced through twin-screw extrusion technology, which is the state-of-the-art in aquatic feed production. The feed costs only ₹400/kg in comparison to imported feed having a price in the range of ₹4,000/kg.

Adoption of Biosecurity and WSSV Incidence

Disease surveillance in the two types of culture systems in West Bengal, shrimp monoculture and traditional bheries over a period of three years (2007–10) indicated that white spot disease (WSD), shell associated problems and gill associated problems were statistically significant among the two culture systems. Newly emerging diseases such as monodon slow growth syndrome was observed in both culture systems, whereas, white faecal disease was observed only in shrimp monoculture system.

WSD outbreak was significantly high where stocking was done by the seed, which has not been tested by PCR. Analysis of data regarding adoption rate of biosecurity protocols either single or in combination in shrimp farms in Chidambaram and Nagapattinam Districts of Tamil Nadu indicated that white spot disease incidence is at a higher rate when protocols of disinfection of implements and reservoir ponds, were not followed.

Diagnostics for Monodon Slow Growth Syndrome

One of the emerging shrimp diseases is the monodon slow growth syndrome (MSGS) with abnormal slow growth and marked size differences among individuals. An improved diagnostic nested RT-PCR with custom designed primer targeting RdRp gene of Laem-Singh virus (LSNV), which has been implicated in monodon slow growth syndrome (MSGS) was developed. Screening of farmed and wild broodstock samples with this improved diagnostic PCR showed a high prevalence of LSNV. High prevalence in broodstock could lead to the spread of this virus by vertical transmission.

4

Husbandry, Housing, and Biosecurity: Facilities and Environment

Proper management is essential for the well-being of the animals, the validity and effectiveness of research and teaching activities, and the health and safety of animal care personnel. Sound animal husbandry programmes provide systems of care that permit the animals to grow, mature, reproduce, and be healthy. Specific operating procedures depend on many factors that are unique to individual institutions. Well-trained and properly motivated personnel can often achieve high quality animal care with less than ideal physical plants and equipment.

ENVIRONMENTAL REQUIREMENTS AND STRESS

Domestic animals are relatively adaptable to a wide range of environments. Domestication is a continuing process. Genetic strains of animals selected for growth or reproduction in different environments under varying degrees of control are used currently for much of the production of livestock and poultry. These strains of animals are sometimes very different from the breeds or strains from which they were originally derived. Agricultural animals may be kept in extensive environments where they reside in large areas outdoors.

They may also be kept in intensive environments where they are confined to an area that would not sustain them were the environment not controlled and where food, water, and other needs must be provided to them. Individual animals may be moved during their lives from extensive to intensive systems or *vice-versa*. Species requirements for domesticated animals are thus variable and depend both on the genetic background of the animals and their prior experience.

CRITERIA OF WELL-BEING

Various criteria have been proposed to identify inappropriate management and housing conditions for agricultural animals. For example, in poultry, significant feather loss that is not associated with natural mating or natural molting is widely accepted as an indication that birds are experiencing stressful conditions. More sophisticated measures of stress are not necessarily superior and may even yield confusing results and lead to inaccurate conclusions.

For instance, plasma corticosteroid concentrations of hens residing in spacious floor pens may be similar to those in high-density cages, even though other criteria may indicate that the caged hens are adversely affected by their environment. During stressful social situations, resistance to virus-induced diseases may be depressed, but resistance to bacterial infections and parasites may be increased. Some researchers have placed emphasis on behavioural criteria of well-being, although others have pointed out the difficulties of interpretation involved.

In the same way, some researchers have suggested that depressed performance of individuals, independent of economic considerations, is a relatively sensitive reflector of chronic stressors, but Hill was less convinced using the same parameters. Animal well-being has both physical and psychological components. No single objective measurement exists that can be used to evaluate the level of well-being associated with a particular system of agricultural animal production. There is consensus, however, that multiple integrated indicators provide the best means of assessing well-being.

Indicators in 4 categories ries are generally advocated:

1. Behaviour patterns;
2. Pathological and immunological traits;
3. Physiological and biochemical characteristics; and
4. Reproductive and productive performance of the individual animal.

A judgement as to the balance of evidence provided by these indicators has been used, when available, as the basis for the recommendations in this guide. D.C. Hardwick postulated and Duncan developed the idea that an acceptable level of animal welfare exists over a range of conditions provided by a variety of agricultural production systems, not under just one ideal set of circumstances.

Improvements in certain environments may increase animal well-being somewhat, but any point in the range would still be considered acceptable with respect to animal welfare. Good management and a high standard of stockmanship are important in determining the acceptability of a particular production system and should be emphasized in agricultural animal research and teaching facilities.

MACROENVIRONMENT AND MICROENVIRONMENT

Animal well-being is a function of many environmental variables, including physical surroundings, nutritional intake, and social and biological interactions. Environmental conditions should be such that stress, illness, mortality, injury, and behavioural problems are minimized. Particular components of the environment that need to be taken into account include temperature, humidity, light, air quality, space, social interactions, microbe concentrations, noise, vermin and predators, nutritional factors, and water.

Environmental Enrichment for further information Physical conditions in the room, house, barn, or outside environment constitute the macroenvironment; the micro-environment includes the immediate physical and biological surroundings. Different microenvironments may exist within the same macroenvironment. Both microenvironment and macroenvironment should be appropriate for the genetic background and age of the animals and the purpose for which they are being used.

Domestic animals readily adapt to a wide range of environments, but some genetic strains have specific needs of which the scientist should be aware and for which accommodation should be made. Even in relatively moderate climatic regions, weather events such as floods, winter storms, and summer heat waves may require that animals have access to shelter. If trees or geographic features do not provide enough protection, artificial shelters and windbreaks or sunshades should be provided.

GENETIC DIFFERENCES

Some strains of agricultural animals may have requirements that differ substantially from those of other stocks of the same species. Some strains of pigs, for example, are particularly susceptible to stress because they carry a gene that causes malignant hyperthermia when they experience even mild stress. Transgenic animals may also have special needs for husbandry and care. Practices to ensure the well-being of special strains should be established independently of those made for the species in general.

SPACE REQUIREMENTS

Floor area is only one of the components that determine the space requirements of an animal. Enclosure shape, floor type, ceiling height, location and dimensions of feeders and waterers, features inside the enclosure, and other physical and social elements affect the amount of space sensed, perceived, and used by the animals in intensive management systems.

When possible, animals in stanchions, cages, crates, or stalls should be allowed to view one another, animal care personnel, and other activities where this would not interfere with research or teaching objectives. Determination of area requirements for domestic animals should be based on body size, head height, stage of life cycle, behaviour, health, and weather conditions.

All area recommendations in this guide refer to the animal zone. Unless experimental or welfare considerations dictate otherwise, space should be sufficient for normal postural adjustments, including standing, lying, resting, self-grooming, eating, drinking, and eliminating feces and urine. When animals are crowded, body weight gain and other performance traits may be depressed and the animals may show altered levels of aggressive behaviour.

TEMPERATURE, WATER VAPOUR PRESSURE, AND VENTILATION

Air temperature, water vapour pressure, and air velocity are some of the most important factors in the physical environment of agricultural animals. In addition, factors related to animal health and genetics affect the thermal balance of animals and thus their behaviour, metabolism, and performance. Most agricultural animals are quite adaptable to the wide range of thermal environments that are typically found in the natural outdoor surroundings of various climatic regions of the continental United States.

The range of environmental temperatures over which animals use the minimum amount of metabolizable dietary energy to control body temperature is termed the thermoneutral zone. Homeothermic metabolic responses are not needed within this zone. Temperature and vapour pressure ranges vary widely among geographic locations. The long-term well-being of an animal is not necessarily compromised each time it experiences cold or heat stress.

However, the overall efficiency of metabolizable energy use for productive purposes is generally lower outside the thermoneutral zone than it is within the zone. The preferred thermal conditions for agricultural animals lie within the range of nominal performance losses. Actual effective environmental temperature may be temporarily cooler or warmer than the preferred temperature without compromising either the overall well-being or the productive efficiency of the animals.

Evaluation of thermoregulation or of heat production, dissipation, and storage can serve as an indicator of well-being in relation to thermal environments. The thermal environment that animals actually experience represents the combined effects of several variables, including air temperature, vapour pressure, air speed, surrounding surface temperatures, insulative effects of the surroundings, and the age, sex, weight, infectious status, transgenic modification status, adaptation status, activity level, posture, stage of production, body condition, and dietary regimen of the animal.

To overcome shortcomings of using ambient temperature as the only indicator of animal comfort, thermal indices have been developed to better characterize the influence of multiple environmental variables on the animal. The temperature-humidity index (THI), first proposed by Thom has been extensively applied for moderate to hot conditions, even with recognized limitations related to airspeed and radiation heat loads.

At the present time, the THI has become the de facto standard for classifying thermal environments in many animal studies and selection of management practices during seasons other than winter. The THI has further been used as the basis for the Livestock Weather Safety Index to describe categories of heat stress associated with hotweather conditions for livestock exposed to extreme conditions. Categories in the LWSI are alert, danger, and emergency. Additionally, THI between 70 and 74 is an indication to producers that they need to be aware that the potential for heat stress in livestock exists.

The index {wind chill temperature index (°C) = 13.12 + (0.6215 × AT)-[11.37 × (WSPD)0.16] + [0.3965 × AT × (WSPD)0.16]}, where AT = air temperature, °C, and WSPD = wind speed, m/s, is a physiological based model and accounts for inherent errors in the earlier wind chill index (WCI), which was not based on heat transfer properties of body tissues.

However, the old WCI closely mimicked heat loss and equivalent temperature equations reported by Ames and Insley for sheep and cattle. Equations developed by Ames and Insley accounted for heat transfer through pelts and hides sections of previously harvested animals; however, they did not account for fat cover and other regulatory processes utilized in mitigating cold stress. In addition, body heat loss due to wind will be proportional to the surface area exposed and not the entire surface area of the body.

This error was also inherent in the old WCI. A ventilation system removes heat, water vapour, and air pollutants from an enclosed animal facility at the same time that it introduces fresh air. Adequate ventilation is a major consideration in prevention of respiratory and other diseases. Where temperature control is critical, cooling or heating may be required to supplement the ventilation system. For certain research projects, filtration or air conditioning may be needed as well.

Typically, ventilation is the primary means of maintaining the desired air temperature and water vapour pressure conditions in the animal microenvironment. The amount of ventilation needed depends on the size, number, type, age, and dietary regimen of the animals, the waste management system, and atmospheric conditions. Equipment and husbandry practices that affect heat and water vapour loads inside the animal house also should be considered in the design and operation of the ventilation system.

Ventilation rates in enclosed facilities should increase from a cold-season minimum to a hot-season maximum. It is important to recognize the approximately 10-fold increase in ventilation rate from winter to summer that is required in a typical livestock or poultry house. Because the animals themselves are the major source of water vapour, heat, and odorous matter, ventilation rate calculated on the basis of animal mass is more accurate than that based on airexchange rate guidelines. Relative humidity is ordinarily the parameter used to manage the air moisture content.

Hot weather ventilation rates should be sufficiently high to maintain the relative humidity below 80 per cent in an enclosed animal house except for situations in which high relative humidity does not cause animal health concerns. Conversely, ventilation rate during cold weather should be sufficiently low to ensure that the relative humidity does not fall to a level that causes animal health concerns, unless needs for air quality or condensation control necessitate a higher rate.

Atmospheric humidity does not ordinarily become a significant factor in effective environmental temperature until the air temperature approaches the temperature of the animal's surface, in which case the animal will depend almost entirely on evaporative heat loss to maintain thermal equilibrium with the environment. The use of fans to promote air movement can be beneficial during hot weather if there is too little natural air movement.

Direct wetting is effective in decreasing heat stress on cattle and pigs; however, it can cause the death of poultry. Wetting is best accomplished by water sprinkled or dripped directly on the animals. Misters and evaporative coolers specifically designed to reduce air dry-bulb temperature are also used to reduce heat stress on agricultural animals. Correctly designed and maintained sunshades protect animals from heat stress by reducing solar radiation load. Trees, if available, are ideal sunshades. Artificial, roofed shades are acceptable.

Mechanical ventilation requires proper design and operation of both air inlets and fans for proper distribution and mixing of the air and thus for creating uniform conditions throughout the animal living space. Mechanical ventilation, with fans creating static pressure differences between inside and outside the house, brings in fresh air and exhausts air that has picked up heat, water vapour, and air pollutants while passing through the building.

Mechanical ventilation, if properly designed, provides better control of air exchange for enclosed, insulated animal houses in colder climates than does natural ventilation. The effectiveness of natural ventilation in cold climates will depend on the design and orientation of the enclosure, as well as the species and number of animals housed and the stage of their life cycle. Natural ventilation uses thermal buoyancy and wind currents to vent air through openings in outside walls or at the ridge of the building.

Natural ventilation is especially effective for cold animal houses in moderate climates; however, insulated walls, ceilings, and floors are often recommended to minimize condensation. The air exchange rate needed to remove the water vapour generated by animals and evaporation of water from environmental surfaces often brings air temperature inside such houses down to values near those outdoors.

If waterers and water pipes are protected from freezing, the practical low operating temperature is the point at which manure freezes, although this temperature would be too cold for some species or stages of the life cycle. Automatic curtains or vent panels, insulated ceilings, and circulating fans help to regulate and enhance natural ventilation systems.

During cold weather, ventilation in houses for neo-natal animals should maintain acceptable air quality in terms of water vapour and other pollutants without chilling the animals. Air speed should be less than 0.25 m/s past very young animals. There should be no drafts on young poultry or pigs. During hot, warm, or cool atmospheric conditions, ventilation of animal houses should maintain the thermal comfort of the animal to the extent possible.

Ideally, the ventilation rate should be high enough to prevent indoor temperature from exceeding outdoor temperature by more than 3°C when the atmospheric temperature is above 32°C for small animals and above 25°C for larger ones. In arid and semi-arid regions where the potential for evaporative heat loss is great, air temperature may peak at over 43°C for 1 or 2 d or longer without affecting animal well-being if animals have been acclimatized by chronic exposure.

Ventilation system design should be based on building construction and the rates of water vapour and heat production of the animals housed. The frame of reference is the animal microenvironment. For example, the outdoor calf hutch is a popular accommodation for dairy replacement heifer calves in most parts of the continental United States. Although the hutch provides a cold microenvironment for calves during winter in northern latitudes, the calf is nonetheless comfortable if cared for correctly.

In closed houses during hot periods, additional ventilation capacity may be necessary. In enclosed animal houses, both environmental temperature and air quality depend on the continuous functioning of the ventilation system. An automatic warning system is desirable to alert animal care and security personnel to power failures and out-of-tolerance environmental conditions and consideration should be given to having an on-site generator for emergency use.

The relative air pressures between animal areas and service areas of a building housing animals should be considered when the ventilation system is designed to minimize the introduction of airborne disease agents or air pollutants into the service area. Advice of a qualified agricultural engineer or other specialist should be sought for the design of and operating recommendations for ventilation equipment.

AIR QUALITY

Air quality refers to the nature of the air with respect to its effects on the health and well-being of animals and the humans who work with them. Air quality is typically defined in terms of the air content of certain gases, particulates, and liquid aerosols, including those carrying microbes of various sorts. Good ventilation, waste management, and husbandry usually result in acceptable air quality. Ammonia, hydrogen sulfide, carbon monoxide, and methane are the pollutant gases of most concern in animal facilities.

In addition, OSHA has established allowable exposure levels for human workers with 8h of exposure daily to these gases. The concentration of ammonia to which animals are exposed ideally should be less than 10 ppm and should not exceed 25 ppm, but a temporary excess should not adversely affect animal health. Comparable concentrations for hydrogen sulfide are 10 and 50 ppm, respectively. The concentration of carbon monoxide in the air breathed by animals should not exceed 150 ppm, and methane should not exceed 50,000 ppm.

Special ventilation is required when underfloor waste pits are emptied because of the potentially lethal hazards to animals and humans from the hydrogen sulfide and methane gases that are released. Many factors affect airborne dust concentration, including relative humidity, animal activity, air velocity, and type of feed. Dust concentration is lower at higher relative humidities. High animal activity and air velocities stir up more particles and keep them suspended longer.

Fat or oil added to feed reduces dust generation. Microbes and pollutant gases may attach to airborne dust particles. The allowable dust levels specified by OSHA are based on exposure of human workers for 8 h daily without facemasks; allowable dust levels are 5 mg/m3 for respirable dust and 15 mg/m3 for total dust.

Although animals can tolerate higher levels of inert dust with no discernible detriment to their health or well-being, the concentration of dust in animal house air should be minimized. Concentrations of microbes in the air should be minimized. Dust and vapour pressure should be controlled. The ventilation system should preclude the mixing of air from infected microenvironments with that from microenvironments of uninfected animals.

LIGHTING

Lighting should be diffused evenly throughout an animal facility. Illumination should be sufficient to aid in maintaining good husbandry practices and to allow adequate inspection of animals, maintenance of the wellbeing of the animals, and safe working conditions for personnel. Guidelines are available for lighting systems in animal facilities.

Although successful light management schemes are used routinely in various animal industries to support reproductive and productive performance, precise lighting requirements for the maintenance of good health and physiological stability are not known for most animals. However, animals should be provided with both light and dark periods during a 24-h cycle unless the protocol requires otherwise.

Red or dim light may be used if necessary to control vices such as feather-pecking in poultry and tail-biting in livestock. Provision of variable-intensity controls and regular maintenance of light fixtures helps to ensure light intensities that are consistent with energy conservation and the needs of animals, as well as providing adequate illumination for personnel working in animal rooms. A time-controlled lighting system may be desirable or necessary to provide a diurnal lighting cycle. Timers should be checked periodically to ensure their proper operation.

EXCRETA MANAGEMENT AND SANITATION

A complete excreta management system is necessary for any intensive animal facility.

The goals of this system are as follows:

- To maintain acceptable levels of worker health and animal health and production through clean facilities;
- To prevent pollution of water, soil, and air;
- To minimize generation of odours and dust;
- To minimize vermin and parasites;
- To meet sanitary inspection requirements; and
- To comply with local, state, and federal laws, regulations, and policies.

The planning and design of livestock excreta management facilities and equipment are discussed by MWPS. A plan should be followed to ensure that the animals are kept reasonably dry and clean and are provided with comfortable, healthful surroundings. Good sanitation is essential in intensive animal facilities, and principles of good sanitation should be understood by animal care personnel and professional staff.

Different levels of sanitation may be appropriate under different circumstances, depending on whether manure packs, pits, outdoor mounds, dirt floors, or other types of excreta management and housing systems are being used. In some instances, animals may be intentionally exposed to excreta to enhance immunity. A written plan should be developed and implemented for the sanitation of each facility housing agricultural animals.

Building interiors, corridors, storage spaces, anterooms, and other areas should be cleaned regularly and disinfected appropriately. Waste containers should be emptied frequently, and implements should be cleaned frequently. It is good practice to use disposable liners and to wash containers regularly. Animals can harbour microbes that can be pathogenic to humans and other species. Hence, manure should be removed regularly unless a deep litter system or a builtup manure pack is being employed, and there should be a practical programme of effective disinfection to minimize pathogens in the environment. For terminal cleaning, all organic debris should be removed from equipment and from floor, wall, and ceiling surfaces.

If sanitation depends on heat for effectiveness, the cleaning equipment should be able to supply water that is at least 82°C. When chemical disinfection is used, the temperature of wash water may be cooler. If no machine is available, surfaces and equipment may be washed by hand with appropriate detergents and disinfectants and with vigourous scrubbing.

Health and performance of animals can be affected by the time interval between successive occupations of intensive facilities. Complete disinfection of such quarters during the unoccupied phase of an all-in, all-out regimen of facility management is effective for disease management in some situations. Programmes of pasture-to-crop rotation for periodically resting the pasture and programmes that permit grazing by other animal species can aid in the control of soilborne diseases and parasites.

Spreading of manure on pastures as fertilizer is a sound and acceptable management practice but may spread toxic agents and infectious pathogens. Caution should be exercised with manure of animals infected with known pathogens, and other methods of waste disposal should be considered. Animal health programmes should stipulate storage, handling, and use criteria for chemicals designed to inactivate infectious microbes and parasites.

There should be information about prevention, immunization, treatment, and testing procedures for specific infectious diseases endemic in the region. Where serious pathogens have been identified, the immediate environment may need to be disinfected as part of a preventive programme. Elimination of moist and muddy areas in pastures may not be possible, but prolonged destocking is an available option.

Drylot facilities may need to be scraped and refilled with uncontaminated materials. Thorough cleaning of animal housing facilities may be followed by disinfection.

Selection of disinfection agents should be based on knowledge of potential pathogens and their susceptibilities to the respective agents. Some means for sterilizing equipment and supplies is essential when certain pathogenic microbes are present and for some specialized facilities and animal colonies. Except in special cases, routine sterilization of equipment, feed, and bedding is not necessary if clean materials from reliable sources are used. In areas where hazardous biological, chemical, or physical agents are being used, a system for monitoring equipment should be implemented.

FEED AND WATER

Animals must be provided with feed and water in a consistent manner, on a regular schedule, in accordance with the requirements established for each species by the NRC and as recommended for the geographic area. When exceptions are required by an experimental or instructional protocol, these must be justified in the protocol and may require approval by the Institutional Animal Care and Use Committee (IACUC). Feeders and waterers must be designed and situated to allow easy access without undue competition.

Sufficient water must be available to meet the animals' daily needs under all environmental conditions. Water troughs, bowls, or other delivery devices must be cleaned as needed to ensure adequate intake and to prevent transmission of microbial-or contaminantassociated disease. Non-municipal water sources should be periodically tested for quality by an approved agency or laboratory. Large supplies of feed should be stored in appropriate, designated areas.

Bulk feed storage containers and feed barrels must be well maintained and the lids kept securely in place to prevent entry of pests, water contamination, and microbial growth. Containers should be cleaned as needed to ensure feed quality. The area around the containers such as the auger boot area should be cleaned regularly. Feed in sacks should be stored off the floor on pallets or racks, and each sack should be labeled with the contents and manufacture date or use-by date.

All feedstuffs should be maintained in such a manner as to prevent contamination by chemicals and/or pests. For example, open feed sacks should be stored in closed containers, and mixing devices and utensils, feed delivery equipment, and feeders/feeding sites should be cleaned regularly to ensure adequate feed intake and prevent transmission of microbial-or contaminant-associated disease. Feed placed in carts or in other delivery devices should be fed promptly or covered to avoid attracting pests.

An effective programme of vermin control should be instituted in feed storage areas. Animal care personnel should routinely inspect feed to identify gross abnormalities such as mold, foreign bodies, or feces; such feed should not be fed until the abnormal components are removed or the feed is determined to be safe. Toxic compounds should be stored in a designated area away from feed and animals to avoid accidental consumption.

SOCIAL ENVIRONMENT

Agricultural animals are social by nature and social isolation is a stressor. Agricultural animals that normally live in herds or flocks under natural conditions that are used in research and teaching should be housed in pairs or groups when possible. Considerations involved in implementing social housing for agricultural animals are discussed by Mench.

If social housing is not feasible because of experimental protocols or because of unpreventable injurious aggression among group members, singly housed animals should be provided with some degree of visual, auditory, and olfactory contact with other members of their species. Socialization to humans and regular positive human contact is also beneficial. In some instances, one species can be used as a companion for another species. Temporary isolation is sometimes required for an animal's safety but the animal should be returned to a social setting as soon as possible.

SEPARATION BY SPECIES

Agricultural animals of different species are typically kept in different enclosures to reduce interspecies conflict, meet the husbandry and environmental needs of the animals, and facilitate research and teaching. However, some research protocols or curricula require species to be co-housed. Facility design and husbandry practices influence whether this can be accomplished in a manner that assures the well-being of the animals.

Mixing of compatible species can often be accomplished more easily in extensive production situations than in intensive housing situations. Some species can carry subclinical or latent infections that can be transmitted to other species that are housed in close proximity, causing clinical disease or mortality. Therefore, a qualified veterinarian or scientist should recommend appropriate health and biosecurity practices if species are to be co-housed.

SEPARATION BY SOURCE OR AGE

Animals obtained from different sources often differ in microbiological status. It is usually desirable to keep these animals separated, at least until microbiologic status is determined or steps are taken to protect against disease transmission. Separation of animals of different ages may also be advisable to reduce disease transmission and control social interactions. Placing animals in groups of similar age or size may allow more uniform access to feed and reduce injuries.

All-in, all-out schemes are examples of age-group separation that are designed to minimize disease risk. However, mixed-group housing is acceptable if disease risk is low, husbandry practices are good, and social interaction is acceptable or necessary. A qualified veterinarian and animal facility manager should work together to devise housing configurations and husbandry practices that assure animal health and well-being while also meeting research and teaching goals.

HUSBANDRY

ANIMAL CARE PERSONNEL

The principal scientist or animal management supervisor should make all animal care personnel aware of their responsibilities during both normal work hours and emergencies. A programme of special husbandry procedures in case of an emergency should be developed. It is the reserach facility management's responsibility to ensure that personnel caring for agricultural animals used for research or teaching are appropriately qualified or trained. This responsibility may be delegated to an IACUC. Qualification by experience and training must be documented.

The animal facility manager must ensure that all animal care personnel are aware of their responsibilities during and outside normal work hours. Protocols for emergency care must be developed and made available to all personnel.

OBSERVATION

Animals in intensive accommodations should be observed and cared for daily by trained and experienced caretakers. Illumination must be adequate to facilitate inspection. In some circumstances, more frequent observation or care may be needed. Under extensive conditions, such as range or pasture, observations should be frequent enough to detect illness or injury in a timely fashion, recognize the need for emergency action, and ensure adequate availability of feed and water.

A disaster plan must be developed for observing animals and providing care during emergency weather or health situations. Regardless of accommodations, animal observations should be documented and husbandry or health concerns reported to the animal facility manager or attending veterinarian as appropriate.

EMERGENCY, WEEKEND, AND HOLIDAY CARE

There must be a means for rapid communication in case of an emergency. In emergencies, facility security and fire personnel must be able to contact staff members responsible for the care of agricultural animals. Names and contact information for those individuals should be posted prominently in the animal facility and provided to the security department or telephone center. If posting names and contact information poses privacy or security issues, a contact number for a security or command center should be used instead.

The institution must ensure that emergency services can be contacted at any time by staff members. The institution must assure continuity of daily animal care, to encompass weekends, holidays, unexpected absences of assigned personnel, and emergency situations. Staff assigned to weekends and holidays must be qualified to perform assigned duties. Cross-training of staff and establishment of standard operating procedures is encouraged to assure consistent, high-quality care.

Emergency veterinary care must be readily available after daily work hours, on weekends, and on holidays. In the event that weather conditions or natural disasters make feeding temporarily impossible, every attempt should be made to provide animals with a continuous supply of water. Absence of feed for up to 48 h should not seriously endanger the health of normal, well-nourished juvenile or adult cattle, sheep, goats, horses, poultry, or swine. Feed should be provided within 24 h to very young animals that are not nursing their dams.

EMERGENCY PLANS

A site-specific emergency plan must be developed to care for agricultural animals that are used for research and teaching. The goal for a plan should be to provide proper management and care for the animals regardless of the conditions. However, some conditions may be so unusual and extreme that it will not be possible to provide immediate care for the animals and to simultaneously ensure employee safety.

Thus, emergency plans should define proper animal management and care and parameters to ensure employee safety. Emergency plans should name employees or positions that are considered essential for providing proper animal management and care. Those employees should a priori understand that responding to emergencies is a condition of employment and that they will be held accountable should they fail to care properly for the animals.

Plans should focus on emergencies that are most likely to occur in the specific geographic area or the research or teaching facility. Emergency plans should include animal evacuation plans specific to the research or teaching facility and actions to be taken if transportation is interrupted.

ANIMAL IDENTIFICATION AND RECORDS

Animals should be permanently identified by a method that can be easily read. Identification of individual animals is desirable, but, in some circumstances, it is acceptable to identify animals by group, cage, or pen. Individual birds may be wing-banded or leg-banded. Ear-notching, ear tattooing, electronic transponders, and branding may be used for individual identification of other species, and each has its advantages and disadvantages. Ear notches and tattoos are permanent and effective, but notching constitutes elective surgery and tattoos generally cannot be read without restraining animals.

Electronic transponders require special sensor units or stations, but should be considered when possible. Cattle and horses are most consistently identified using freeze-branding on the hip, shoulder, rear leg, or side. In addition, when freeze branding is used on more than one breed of horse, branding is performed under the mane. Some states require that cattle be permanently identified by branding with a hot iron; however, this procedure is more stressful than freeze-branding.

Ear and neck chain tags, although readable at some distance, can become lost and are therefore not necessarily permanent. In addition, neck chains and straps should be avoided in situations in which the animal could become entangled in a fence, rock outcropping, or other feature of the environment. Any associated pain and distress should be considered when determining the method of identification. In some cases, it may be necessary to identify animals in multiple ways. Individual records are needed for most animals.

These records should include information about the animal, its source and location, its productivity, its reproductive performance, protocols the animal is assigned to, and its ultimate disposition. Records for individual animals or groups should also include dates of vaccination, parasite control measures used, blood testing dates and results, and notations as to whether castration, spaying, or other elective procedures have been performed.

Applicable veterinary data to be recorded include dates of examination/treatment, clinical information/diagnosis, names of medications and amounts and routes of administration, descriptions of surgical procedures, and resolution of surgical procedures or illnesses. Principal scientists or animal facility managers may wish to record nutritional information. Research protocols often dictate that additional information be recorded.

VERMIN CONTROL

Programmes should be instituted to control infestation of animal facilities by vermin. The most effective control in facilities prevents entry of vermin into the facility by screening openings and ceilings; sealing cracks; eliminating vermin breeding, roosting, and refuge sites; and limiting access of vermin to feed supplies and water sources. Building openings should be screened with 1.3-cm mesh, and ceilings with ridge vents should be screened with 1.9-cm mesh to minimize rodent and bird entry. Smaller mesh sizes are recommended where they will not interfere with airflow.

Mesh may need to be installed along foundations below ground level, especially with wood foundations. Pesticides should be used only as approved. Particular caution should be exercised with respect to residues in feedstuffs, which could injure animals and eventually pass into the meat, milk, or eggs. Pesticides should be used in or around animal facilities only when necessary, only with the approval of the scientist whose animals will be exposed to them, and with special care.

A pesticide applicator or a commercial service may be used. In some regions, wildlife and stray cats and dogs may spread zoonotic diseases, including rabies, to agricultural animals. In highrisk locations, institutions should implement an educational programme that includes training scientific and animal care personnel to recognize the signs of rabies in both wildlife and agricultural species and to handle and report potentially rabid animals.

Inoculation may be advisable for humans who may come into contact with animals in regions where rabies is endemic. Many agricultural institutions keep cats for pest-control purposes. Although the use of free-roaming cats is a traditional form of pest control for agricultural facilities, cats may limit the ability for baiting and may present hygiene or accident risks or serve as disease vectors. However, when cats are present, proper veterinary care and oversight should be provided to these animals. Veterinary care should include vaccinations, parasite control, and neutering.

STANDARD AGRICULTURAL PRACTICES

Sometimes procedures that result in temporary distress and even some pain are necessary to sustain the long-term welfare of animals or their handlers. These practices include comb-, toe-, and beak-trimming of chickens; bill-trimming of ducks; toenail removal, beak-trimming, and snood removal of turkeys; dehorning and hoof-trimming of cattle; taildocking and shearing of sheep; tail-docking, neo-natal teeth-clipping, hoof-trimming, and tusk-cutting of swine; and castration of males and spaying of females in some species.

Some of these procedures reduce injuries to humans and other animals. Castration, for example, reduces the chances of aggression against other animals. Bulls and boars also cause many serious injuries to humans. Standard agricultural practices that are likely to cause pain should be reviewed and approved by the IACUC. The development and implementation of alternative procedures less likely to cause pain or distress are encouraged. Overall, best practices for pain prevention and control should be followed.

SICK, INJURED, AND DEAD ANIMALS

Sick and injured animals should be segregated from the main group when feasible, observed thoroughly at least once daily, and provided veterinary care as appropriate. Incurably ill or injured animals in chronic pain or distress should be humanely killed as soon as they are diagnosed as such.

Dead animals are potential sources of infection. Their disposal should be accomplished promptly by a commercial rendering service or other appropriate means and according to applicable ordinances and regulations. Postmortem examination of fresh or wellpreserved animals may provide important animal health information and aid in preventing further losses. When warranted and feasible, waste and bedding that have been removed from facilities occupied by an animal that has died should be moved to an area that is inaccessible to other animals.

SPECIAL CONSIDERATIONS

NOISE

Noise from animals and animal care activities is inherent in the operation of any animal facility. Although differences exist in perceived loudness of the same sound occupational noise limitations have been established for workers, and employees should be provided appropriate hearing protection and monitored for their effects.

Noise ordinarily experienced in agricultural facilities generally appears to have little permanent effect on the performance of agricultural animals, although Algers and Jensen found that continuous fan noise disrupted suckling of pigs. Sudden loud noises have also been reported to cause hysteria in various strains of chickens.

METABOLISM STALLS AND OTHER INTENSIVE PROCEDURES

Animals that are subjected to intensive procedures requiring prolonged restraint, frequent sampling, or other procedures experience less stress if they are trained to cooperate voluntarily with the procedure. Cattle, pigs, and other animals can be trained with food rewards to accept and cooperate with various procedures, such as jugular venipuncture. Many studies of the nutrition and physiology of agricultural animals use a specialized piece of equipment, the metabolism stall. Successful designs have been reported for various species.

These stalls give animal research and care personnel easy access to the animal and its excreta. The degree of restraint of animals housed in metabolism stalls is substantially different from that of other methods that restrict mobility. Animals in metabolism stalls are often held by a head gate or neck tether and are restricted in their lateral and longitudinal mobility. These differences may exacerbate the effects of restriction on animals housed in metabolism stalls. Metabolism stalls should be used only for approved studies, not for the purpose of routine housing.

Researchers should consider appropriate alternatives to metabolism stalls if such alternatives are available. There should be a sufficient preconditioning period to ensure adequate adjustment and comfort of the animal to the metabolism stall before sample collection starts. The length of the preconditioning period should be subject to approval of the IACUC. At least enough space should be provided in the metabolism stall for the animal to rise and lie down normally. When possible, metabolism stalls should be positioned so that the animal is in visual, auditory, and olfactory contact with conspecific animals to minimize the effects of social isolation.

Thermal requirements of animals may be affected when they are placed in metabolism stalls. For example, the lower critical environmental temperature of an animal held individually in a metabolism stall is higher than when residing in a group because the single animal cannot obtain the heat-conserving benefits of huddling with group-mates.

Animals in metabolism stalls should be observed more frequently than those in other environments, and particular attention should be paid to changes in behaviour and appetite and the condition of skin, feet, and legs. The length of time an animal may remain in a metabolism stall before removal for exercise should be based on professional judgement and experience and be subject to approval by the IACUC. The species and the degree of restraint imposed by particular stall types should be taken into consideration in making such judgements.

BIOSECURITY

The term biosecurity in an agricultural setting has historically been defined as the security measures taken to prevent the unintentional transfer of pathogenic organisms and subsequent infection of production animals by humans, vermin, or other means. Biosecurity is also applied in the same context to agricultural animals used in the field of agricultural research, teaching, and testing. With the advent of bioterrorism and the designation of select agents, the term biosecurity has acquired new definitions, depending on the field to which it is applied.

Biosecurity is now used to define national and local policies and procedures that address the protection of food and water supplies from intentional contamination and is additionally used to define measures required to maintain security and accountability of select agents and toxins. It is important to understand these concepts when using the term and to clarify that in this section we are using the term biosecurity in the context of preventing the unintentional transfer of pathogens to animals and humans through appropriate facility design, training, and precautions.

For example, personnel working in swine and poultry facilities should be immunized against influenza and receive training related to potential cross-contamination of agents between animals and humans. The USDA has published voluntary guidelines and a checklist as a resource to help the agricultural producer reduce security risks at the farm level. This publication is designed to prevent both intentional and unintentional introduction of pathogens at the farm level. A list of references and resources is also provided in this document on a variety of farm biosecurity issues. Other sources of information include reviews of biosecurity basics and good management practices for preventing infectious diseases and biosecurity of feedstuffs. All of these publications offer information and suggestions that could be evaluated for their impact on the design of an animal facility. It is essential that the agricultural animal care staff maintain a high standard of biosecurity to protect the animals from pathogenic organisms that can be transferred by humans. Good biosecurity begins with personal cleanliness.

Showering or washing facilities and supplies should be provided, and personnel should change their clothing as often as necessary to maintain personal hygiene. Disposable gear such as gloves, masks, coats, coveralls, and shoe covers may be required under some circumstances. Personnel should not leave the work place in protective clothing that has been worn while working with animals. Personnel should not be permitted to eat, drink, apply cosmetics, or use tobacco in animal facilities.

Visitors should be limited as appropriate, and institutions should implement appropriate precautions to protect the safety and well-being of the visitors and the animals. Preventing the introduction of disease agents is a continuous challenge, particularly when teaching and research facilities allow public access. Herd health and sanitation programmes should be in place to minimize exposure to pathogens.

Animal care personnel in research and teaching facilities should not be in contact with livestock elsewhere unless strict biosecurity precautions are followed. To reduce inter-building transmission of pathogenic microorganisms, careful attention should be given to traffic patterns of inter-building personnel and disease organisms in feed and transport vehicles.

Barriers to microorganism transmission should be considered for personnel who move between houses, including showering in, changing clothes, and the use of disinfectant footbaths as personnel move between rooms and buildings. Establishing a barrier between animals and visitors requires visitors to do some or all of the following: shower in/shower out, wear clean footwear, change to on-site clothes, and wear only on-site clothes. In addition, if personnel need to go back and forth between different phases of production, it is critical that they work from clean to dirty phases of the farm.

Boot Cleaning and Disinfection

The use of boot baths can prevent or minimize mechanical transmission of pathogens among groups of pigs. Visible organic material may be removed from boots using water and a brush or specific boot cleaning station. Boots may be disinfected by soaking in a clean bath of an appropriate disinfectant following the manufacturer's guidelines for dilution rate and exposure time. Personnel should step into and scrub their boots in the boot bath upon entry and when leaving the room/facility. It is important to frequently empty, clean, and refill the boot bath to prevent it from being contaminated with organic matter. Disposable boots may be used.

BIOCONTAINMENT

High-consequence livestock pathogens or the vectors responsible for transmission of disease cause high morbidity and mortality, and can have a significant regional, national, and global economic impact. The use of these pathogens in agricultural research brings several challenges when designing and operating an animal facility. The design of this type of facility should strive for flexibility, effective containment of pathogens, and minimizing the risk of exposure to personnel when zoonotic agents are utilized.

The use of agricultural animals in high-consequence livestock pathogen research requires a thorough understanding of a variety of regulatory requirements and the concept of risk assessment. The USDA provides a list of livestock, poultry, and fish pathogens that are classified as "pathogens of veterinary significance" *Biosafety in Microbiological and Biomedical Laboratories*. The use of these pathogens requires facilities to meet specific criteria for design, operation, and containment features, which are described in the BMBL.

For the listed agents, criteria may include utilizing containment levels designated as:

- Animal Biosafety Level (ABSL)-2,
- Enhanced ABSL-3,
- BSL-3-Ag, or
- ABSL-4.

Requirements for BSL-3-Ag facilities must be met when any of the listed pathogens are used in animals and the room housing the animals provides the primary containment. When the studies can be accomplished in smaller species in which animals are housed in primary containment devices, which allows the room to serve as the secondary barrier, then enhanced ABSL-3 requirements can be utilized.

Enhancements to ABSL-3 should be determined on a case-by-case basis, using risk assessment, and in consultation with the Animal and Plant Health Inspection Service (APHIS) of the USDA. In addition to the BMBL, facility design standards have been published by the USDA to guide the design of Animal Research Service (ARS) construction projects and contain useful information on the design of containment facilities for agricultural research.

These standards include information on containment design that addresses hazard classification and choice of containment, containment equipment, and facility design issues for the different levels of biocontainment. Although published to provide guidance for National Institutes of Health (NIH)-funded construction projects and renovations for biomedical research facilities, the *NIH Design and Policy Guidelines* contain useful information on construction of BSL-3 and ABSL-3 facilities. The use of recombinant DNA molecules in agricultural research can introduce additional considerations when designing an animal facility.

Published guidelines provide recommendations for physical and biological containment for recombinant DNA research involving animals. These guidelines include a supplement published in 2006 that provides additional information specific to the use of lentiviral vectors. The Agricultural Bioterrorism Protection Act of 2002 required the propagation of regulations that address the possession, use, and transfer of select agents and toxins that have the potential to pose a severe threat to plants or animals, and their products.

The USDA/APHIS published the implementing regulation covering animals and animal products, which identifies those select agents and toxins that are a threat solely to animals and animal products and overlap agents, or those agents that pose a threat to public health and safety, to animal health, or to animal products. Overlap select agents and toxins are subject to regulation by both APHIS and the Centers for Disease Control and Prevention.

The regulations implemented by both agencies reference the BMBL and the NIH *Guidelines for Research Involving Recombinant DNA Molecules* as sources to consider when developing physical structure and features, and operational and procedural safeguards. Other issues discussed in some of these references may not directly affect containment of pathogens or safety of personnel, but should be considered as they may affect the design of a facility.

For example, the use of select agents requires certain security measures to be in place that restrict access to areas where select agents or toxins are used or stored. This can include laboratories, animal rooms, and storage freezers, resulting in a significant impact on how a research facility is designed. A thorough understanding of the references cited in this section is advised before initiating the design of new biocontainment facilities or renovation of existing facilities to accommodate research with hazardous agents or toxins requiring containment.

5

The Primary Alternative to Conventional Farm Animal Health Care

ANIMAL HEALTH

DISEASE DIAGNOSIS

Successful control of a disease requires accurate diagnosis. This has been greatly improved in recent years through developments in biotechnology. The most recent major development, the finding that it is possible to immortalise individual antibody-producing cells by hybridisation to produce antibodies of a given class, specificity and affinity (*i.e.* monoclonal antibodies) has provided a tool that permits the analysis of virtually any antigenic molecule. The use of monoclonal antibodies has revealed that the failure of vaccines (*e.g.* of rabies) to provide protection in all parts of the world was due to the diversity in the antigenic composition of the causative virus. The (monoclonal antibody) technology is relatively simple and can readily be applied in developing countries. Monoclonal antibodies are currently supplied to developing countries directly or in the form of kits and simple reagents for completion of the tests. For example, kits for rinderpest virus diagnosis used in African countries come in this form.

The ability to generate highly specific antigens by recombinant DNA techniques has made it possible for an increasing number of enzyme-linked immunosorbent assays (ELISA) to have the capacity to differentiate between immune responses generated by vaccination from those due to infection. This has made it possible to overcome one of the major drawbacks of antibody detection tests: the fact that, because antibodies can persist in animals for long periods, their presence may not indicate current infection. ILRAD has developed a technique to overcome this problem in diagnosis of trypanosomiasis.

The parasite antigen detection test uses monoclonal antibodies raised in laboratory mice to capture the parasite antigens which are then revealed by their reaction with a second layer antibody to which is conjugated an easily detectable enzyme. This test reveals current infections and facilitates differentiation between the major trypanosome species. This has important implications for disease control, especially because of the association of different parasite species with different epidemiological and disease circumstances.

The advent of PCR has enhanced the sensitivity of DNA detection tests considerably. For example, PCR used in combination with hybridisation analysis, has been shown to provide a sensitive diagnostic assay to detect bovine leukosis virus. Other diagnostic techniques include nucleic acid hybridisation (NAD) and restriction endonuclease mapping (REM). As has been indicated above, one of the most valuable features of these molecular techniques is their specificity and sensitivity. A good example of the specificity of NAD is its application in distinguishing infections caused by *peste des petite ruminants* (PPR) virus from rinderpest, diseases whose symptoms are clinically identical and which cannot be distinguished antigenically with available serological reagents. This technique also allows comparison of virus isolates from different geographical locations. Detailed overview of biotechnological tools for diagnosis of livestock diseases has been provided by Bourne and Bostock and Robinson and McEvoy.

Other available diagnostic techniques which may have application in small ruminants and/or cattle include: nucleic acid probes (NAP) for heartwater *(Cowdria ruminantium), chlamydia psitacci,* Paratuberculosis and Bluetongue; restriction endonuclease reaction (RENR) for diagnosis of *Corynebacterium pseudotuberculosis*; PCR for characterising subtypes of Bluetongue from different geographical regions; monoclonal antibodies (MAB) for differentiating false positive anti-Brucella titres caused by *Yersinia enterocolitica* and true positive anti-Brucella titres in latent infected animals; and MAB for diagnosis of Toxoplamosis, Pasteurellosis, *Mycoplasma* spp, PPR and Boder Disease.

Table. Tropical Diseases for which Probes and Monoclonal Antibodies (MAB) are Available.

Viral diseases	Bacterial diseases
Rinderpest	Cowdriosis
Peste des petite ruminants	Contagious bovine pleuropneumonia
Bluetongue	Contagious agalactia
African horse sickness	Contagious caprine pleuropneumonia
Foot-and-mouth disease	Anaplasmosis (A. centrale and A. marginale)
Vesicular stomatitis	Haemorrhagic septicemia (only MAB available)

Examples of tropical diseases for which diagnostic tests are available are presented in Table. Teale reviewed the diagnostic techniques that are currently available or which could be developed in the near future.

VACCINES

Conventional means of controlling major livestock diseases include chemotherapy, vector control, vaccination, slaughter of infected stock, and other management practices (including grazing management and controlled stock movements).

Vector control requires continuous application of pesticides. These are often unaffordable to farmers in developing country. Moreover, where these drugs or pesticides are used, resistance by parasites is often encountered and reinfection following administration of drugs against parasitic diseases usually occurs. Additionally, in many cases drugs are not readily available locally. In some cases where they are available, they are ineffective, either because they have been partly preserved or they are not genuine.

Immunisation remains one of the most economical means of preventing specific diseases. An effective vaccine can produce long-lasting immunity. In some cases, vaccination can provide lifetime immunity. Moreover a small number of doses is usually required for protection. Level of infrastructure and logistical support required for a large-scale vaccination programme is such that a successful vaccination campaign can be implemented in remote rural areas. In general, vaccines offer a substantial benefit for comparatively low cost, a primary consideration for developing countries.

Vaccines have conventionally been produced by several methods some of which have become rather static with regard to efficacy, safety, stability and cost. Very effective vaccines against animal diseases such as rinderpest and pig cholera have been in use for more than 20 years and have helped to significantly reduce the incidence of these diseases world-wide.

However, vaccines of questionable efficacy also exist. Impotency, instability, adverse side effects, and reversion of attenuated organisms to wild (disease-causing) forms represent some of the problems. However, research strategies for the development of better, cheaper and safer vaccines are constantly being sought.

Through the use of monoclonal antibodies and recombinant DNA technologies, it is now possible to define and produce immunogenic components much more rapidly. These technologies are increasingly being used to clarify the pathogenetic mechanisms and immune response to microbial diseases. This should lead to the production of more effective vaccines in the future. To date many candidate vaccines have been produced by these techniques. However, only few of these are being produced commercially.

Table summarises some vaccines developed by recombinant DNA technology. There are other vaccines under various stages of development. The following are of particular relevance to small ruminants: recombinant vaccines against Bluetongue and Rift Valley Fever may soon be available for field use; a hybrid virus vaccine against Orf, a thermostable recombinant vaccine against PPR and recombinant vaccines against *Taenia ovis* and Echinococcus; tick gut antigens and *Haemonchus* gut antigens produced by molecular techniques offer ways for prophylaxis against these parasites.

Progress is also being made towards the development of vaccines against Babesiosis and Theileriosis in cattle which may indicate prospects for similar immune-prophylaxis becoming available for sheep and goats as well.

Table. Example of Some Novel Animal Vaccines

Species	Disease	Producer
Avian	Coccidiosis,	Genex and A.H. Robins
	Newcastle virus	Codon and Salsbury labs
Bovine	Papilloma virus	Molecular genetics
	Viral diarrhoea	California Biotechnology
	Brucellosis	Ribi ImmunoChem
	Rinderpest	USDA and University of California, Davis
Swine	Parvovirus	Applied Biotechnology
	Dysentery	Codon
Equine	Influenza, Herpes	California Biotechnology
		Applied Biotechnology
Companion	Canine parvovirus	Applied Biotechnology

Physiology of Lactation and Growth

Recombinant bovine somatotropin (BST) is a genetically engineered synthetic analog of the natural growth hormone. Since the 1970s there have been a number of studies on the effects of BST on milk yield, reproductive performance and health as well as its likely effect on humans who consume such milk. Under good management and feeding, regular BST administration to lactating dairy cows increases milk yield by 15-30 per cent and also increases efficiency of milk production.

BST is now readily available and is already in commercial use in the United States. However, the appropriateness of BST use to increase milk production in the USA is doubtful as the country already has a milk surplus and the public is also concerned about its effects on health. Indeed, the benefit of BST in the USA is perceived to be towards reducing the production costs of large dairy farms, but this could push smaller farmers out of the market. The appropriateness of BST use for developing countries is still a matter of debate. Those supporting its introduction argue that its use in commercial dairy farms could increase the national milk output. Those opposed point out the fact that

BST does not improve milk yield in indigenous non-dairy breeds and that its use on crossbred and exotic dairy cattle will require more feeding, and that provision of adequate nutrition is already a problem for most dairy operations in developing countries. Whether or not BST affects reproductive performance has not yet been conclusively established. Although most reports have indicated non-significant effects of BST on reproductive performance, most such studies have been on single lactations; the effect of BST on lifetime productivity needs to be investigated. Some studies have shown that feed conversion efficiency declines in subsequent lactations.

The study by Phipps *et al.* also indicated a decline in the incremental amount of milk yield from BST application in subsequent lactations. Additionally, the effect of BST on reproductive performance is likely to be more adverse in the presence of higher biotic and abiotic stresses, including nutritional stresses. There is also need to examine the economics of BST application in view of the known association of its use with mastitis and other infections. BST use is thus bound to be associated with increased use of antibiotics and other veterinary drugs.

Thus, in evaluating the potential role of BST in developing countries, one needs to consider not only the possible response levels of the cattle in these countries to BST treatment, but also the cost of BST, the amount and cost of other incremental inputs required for effective use of BST, and the milk prices. Ultimately, the main technical constraint to BST use in developing countries will not only be its cost, but the absence of an efficient delivery system; current use of the technology requires regular injections.

Porcine somatotropin (PST) and recombinant growth hormone stimulatory peptides (*e.g.* growth hormone releasing factor, GRF) along with BST have been shown to increase growth rates by 8-38 per cent in cattle, sheep and pigs. In almost all cases, administration of exogenous growth hormones have been associated with increased carcass protein and reduced carcass fat. Other growth-promoting agents (*e.g.* anabolic steroids and beta agonists) have been shown to have even larger effects but public concerns over the possible residual effects in meats have led to their being banned in most developing countries. The use of anabolic implants is, however, permitted in some countries such as the USA.

AGRICULTURAL ANIMAL HEALTH CARE

Agricultural animal health care involves proper management and husbandry as well as veterinary care. Proper management is essential for the well-being of animals, the validity and effectiveness of research and teaching activities, and the health and safety of animal care personnel.

Sound animal husbandry programmes provide systems of care that permit the animals to grow, mature, reproduce, express some species-specific behaviour, and be healthy. Specific operating procedures depend on factors that are unique to individual institutions. Well-trained and motivated personnel can often achieve high-quality animal care with less than ideal physical plants and equipment.

VETERINARY CARE

ATTENDING VETERINARIAN

The agricultural animal health care programme is the responsibility of the attending veterinarian. The Institute for Laboratory Animal Research (ILAR), National Research Council *Guide for the Care and Use of Laboratory Animals* defines the attending veterinarian as "a veterinarian who has direct or delegated authority" and who "should give research personnel advice that ensures that humane needs are met and are compatible with scientific requirements." Animal Welfare Act regulations and the Public Health Service policy require that the attending veterinarian have the authority to oversee the adequacy of other aspects of animal care and use, including animal husbandry and nutrition, sanitation practices, zoonoses control, and hazard containment.

Research and teaching institutions must provide investigators and instructors with access to a veterinarian who has experience in the care of agricultural animals. The veterinarian can be full-time or part-time and must have authority to ensure that the provisions of the programme are met. The attending veterinarian must be provided access to all research and teaching animals and to any related documents including health care records. The attending veterinarian also must be involved in the development and oversight of the veterinary care programme, as well as in other aspects of animal care and use such as protocol review, establishment of anesthetic and analgesic guidelines, study removal criteria, training of animal users, and responsible conduct of research activities.

Veterinary involvement in these activities helps to ensure animal health and well-being. The attending veterinarian is not required to be the sole provider of veterinary care and can delegate authority to another qualified veterinarian. However, the attending veterinarian must communicate with, and oversee veterinary care provided by, other veterinarians. When necessary, the attending veterinarian should utilize the expertise of other professionals when making determinations about agricultural animal care. Trained non-veterinary staff may administer treatments according to standard operating procedures approved by the attending veterinarian.

PREVENTIVE MEDICINE

Adequate agricultural animal health care in research and teaching involves a written and implemented programme for disease prevention, surveillance, diagnosis, treatment, and endpoint resolution. The objectives of such a programme are to ensure animal health and wellbeing, minimize pain and distress, maintain animal production, prevent zoonoses, provide assistance to investigators on study-related animal health issues, and avoid contaminants or residues in animal products. The programme should include training for animal users regarding animal behaviour, humane restraint, anesthesia, analgesia, surgical and postsurgical care, and euthanasia. A mechanism for direct, frequent, and regular communication must be established among personnel who are responsible for daily animal care and observation, animal users, and the attending veterinarian. This will help ensure that timely and accurate animal health information is effectively communicated.

SICK, INJURED, AND DEAD ANIMALS

Animal care personnel must be trained to recognize signs of illness and injury. In general, sick and injured animals should be segregated from the main group to protect them and the other animals, observed at least once daily, and provided with veterinary care as appropriate.

When animals are separated, a mechanism should be in place to communicate to staff the status of the animals and to ensure proper daily, weekend, holiday, and emergency care. In some circumstances, segregation is not feasible or may disrupt the social hierarchy, cause additional stress to the animal, or adversely affect research. The advantages of segregation should be weighed against its disadvantages, especially for mild illnesses or injuries that can be easily managed. Care should be taken to minimize spread of pathogens from ill animals to healthy animals by observing appropriate biocontainment measures. Incurably ill animals or ill or injured animals with unrelievable pain or distress should be humanely killed as soon as possible.

Unexpected deaths should be reported to the attending veterinarian. Dead animals are potential sources of infection and should be disposed of promptly by a commercial rendering service or other appropriate means following applicable state and local ordinances and regulations. Postmortem examination of fresh or well-preserved animals may provide important animal health information and aid in preventing further losses. When warranted, waste and bedding that have been removed from a site occupied by an animal that has died should be moved to an area that is inaccessible to other animals and the site appropriately disinfected.

MEDICAL RECORDS

An important component of an agricultural animal health programme is maintaining records that can be used to monitor animal health events, both physical and behavioural health events, as well as outcomes and levels of production. Medical records should comply with the American College of Laboratory Animal Medicine (ACLAM) statement on medical records. Group health records may be appropriate for animals that are kept as cohorts, particularly when animals undergo periodic evaluation by means of examining several representative individuals of the group. The institution, under the guidance of the attending veterinarian, should determine the method(s) by which medical records are maintained. Oversight of medical records is the responsibility of the attending veterinarian and the IACUC.

When institutional representatives determine that a medical record should be created, the record typically contains the following information:

- Identification of the animal(s) or group(s);
- Clinical information, such as the animal's behaviour, results of physical examinations, and observed abnormalities, illnesses, and/or injuries;
- Immunizations and other prophylactic treatments and procedures;
- Documentation and interpretation of diagnostic tests;
- Documentation of research interventions;
- Treatments prescribed and administered;
- Clinical response and follow up information;
- Descriptions of surgical procedures, anesthesia, analgesia, and perioperative care;
- Methods used to control pain and distress;
- Documentation of resolution;
- Documentation of euthanasia or other disposition; and
- Necropsy findings if necropsy is indicated.

The record system must be structured so that information is easily collected, gathered, analysed, summarized, and available to the veterinarian, the principal scientist, and the IACUC.

The ACLAM statement on *Medical Records for Animals used in Research Teaching and Testing* suggests that:

- Notations in the medical record should be made by individuals who have administered treatments, or made direct observations or evaluations of the animal(s) or their diagnostic results, or their designee. Individuals typically responsible for making notations in the record include veterinary staff, animal husbandry staff, and research staff. All entries in the record should be dated, indicate the originator of the entry and be legible to someone other than the writer.

ANIMAL PROCUREMENT

When an institution acquires new animals, attention must be paid to applicable international, federal, and state regulations and institutional procedures, particularly those dealing with transportation and animal health. All animals must be obtained and transported legally. The attending veterinarian, in conjunction with the principal scientist, should formulate written procedures to assess the health status of a herd or flock obtained from a vendor before acquiring animals. The institution should develop a mechanism and process of control for animal acquisition that ensures coordination of resources that will preclude the arrival of animals in advance of preparation of adequate housing and appropriate veterinary quarantine procedures. Quality control for vendors and knowledge of the history of purchased animals is part of an adequate institutional veterinary care programme.

Animals of unknown origin or from stockyards should only be used if necessary; such animals may pose significant unknown health risks compared with animals of known origin and therefore should be handled appropriately. Newly acquired animals should undergo a quarantine and acclimation period, including preventive and clinical treatments as appropriate for their health status.

ACCLIMATION AND STABILIZATION

Newly arrived animals require a period of acclimation. Acclimation refers to a stabilization period, before animal use, which permits physiological and behavioural adaptation to the new environment. The attending veterinarian should establish general acclimation guidelines for each species. Any modifications to the general programme should be discussed with the attending veterinarian before animals are shipped. In some cases, animals may require an extended acclimation period because of their history or health status. On the other hand, some studies, such as comparisons of metaphylactic treatments for shipping fever, need to begin as soon as animals arrive. Such exemptions from the acclimation period must be scientifically justified and approved by the Institutional Animal Care and Use Committee (IACUC).

QUARANTINE

Quarantine is the separation of newly received animals from those already in the facility or on the premises until the health of the new animals has been evaluated and found to be acceptable. The attending veterinarian should ensure that quarantine facilities or locations are appropriate and that quarantine procedures are consistent with current veterinary practices and applicable regulations. The quarantine period should be long enough to observe signs of infectious disease or obtain diagnostic evidence of infection status.

Quarantine and testing of animals before introduction is especially important for herds or flocks that have attained specific-pathogen-free status, but these additions should be discouraged. If the health history of newly received animals is unknown, the quarantine programme should be more comprehensive and sufficiently long to allow expression or detection of diseases present in the early incubation stage. Exceptions to quarantine practices should be approved by the attending veterinarian in advance of shipment of the animals.

The attending veterinarian, or skilled personnel under the direction of the attending veterinarian, should perform an initial examination and subsequent daily observations for newly arrived animals. Animals should be observed in quarantine until they are cleared for introduction into a herd or facility.

During the quarantine period, animals should be vaccinated and treated for diseases and parasites as appropriate to protect their health and maintain the health of animals in the home facility. In addition to having adequate quarantine procedures, research facilities and animal use protocols should be designed to minimize the risk of introducing or transmitting disease agents.

BIOTECHNOLOGY AND ANIMAL HEALTH

The potential areas of impact for biotechnology on the control of infectious diseases and improved animal health are manifold. They include the identification of susceptible stock, tests for the rapid and early diagnosis of disease, an understanding of the mechanisms of pathogenesis and the development of safe and effective vaccines or prophylactics. The principal methodologies which will enable these targets to be approached are those of monoclonal-antibody production and recombinant-DNA technology.

The benefits of selecting stock on the basis of resistance to specific infectious agents would appear at present to be restricted to those areas of the world where the importance of a single disease predominates over all others (*e.g.* trypanosomiasis). Whilst the possibility of selecting on an antigen non-specific basis is likely to be of more general relevance, at present the development of vaccination strategies that can effectively control infectious agents is likely to be of greatest significance for improved animal health.

To this end monoclonal antibodies can be used in the development of diagnostic sera and reagents, and the characterization of virulence and adhesion determinants. Once identified such determinants can be isolated in a pure form from crude mixtures with monoclonal immunosorbents. Infectious diseases may be controlled by the prophylactic use of monoclonal antibodies, but this direct effect will be of limited use in animal production, by comparison with their potential use as 'educators' of the immune system in anti-idiotype vaccines.

Recombinant-DNA technology can provide a range of potential immunogens including attenuated or non-pathogenic strains and subunit vaccines. Since the majority of infectious agents are met at mucosal surfaces, the problem that remains is how these vaccines may be delivered so as to provide effective presentation and so stimulate a protective response.

MUCOSAL DEFENCE

The level of challenge to, and the degree of sterility maintained differs markedly between the various mucosal surfaces; from the gastrointestinal and upper respiratory tracts to the lower respiratory tract, reproductive tract and mammary gland which at least in health remain sterile. They are presented with a vast array of different antigens ranging from potential pathogens to relatively harmless dietary components. Together with the skin they provide the interface between the host and its environment.

However, unlike the skin there are other physiological roles (*e.g.* gaseous exchange, absorption of nutrients, etc.) which restrict their ability to keep antigens out. Since the mucosal surfaces provide such a primary point of contact it is at these sites that the products of biotechnology might be expected to provide the most relevant and beneficial effects. For convenience the major part of the present review will be restricted to antigens presented to the gut. The immune system of the gastrointestinal tract is presented with two types of antigens, those capable of causing infection and those, such as dietary antigens, which in themselves impose no direct threat to the animal's health. The type of response stimulated is critical, for whilst failure to respond to potential pathogens may be fatal, an over-reaction to dietary antigens may lead to damaging hypersensitivity responses.

Latter may in turn lead to altered handling of potential pathogens. Therefore, although the most obvious area of impact of biotechnology will be directly on infectious agents, regulating responses to food antigens may also be of significance.

THE RESPONSE TO FOOD ANTIGENS

Three responses have been described that may protect an individual from damaging hypersensitivity reactions to dietary antigens. Immune exclusion mediated by locally produced IgA may reduce the absorption of intact proteins across the gut mucosae and in certain species it has been shown that any antigen that is absorbed may be eliminated via bile. These processes which restrict the access of antigenic material to the immune system are not absolute and the development of an acquired specific hyporesponsive state is probably of greater significance.

Thus orally induced tolerance may actively down-regulate the response to dietary proteins. Before the induction of this 'protected state' all animals pass through a transient phase of hypersensitivity, during which time damaging allergic reactions can be demonstrated. In farm species it has been shown that following weaning, piglets are sensitive to antigens in the post-weaning diet and since these changes occur before the proliferation of *Escherichia coli* it has been postulated that response to food proteins may be an important predisposing factor in the aetiology of post-weaning diarrhoea.

The importance of gut flora and bacterial lipopolysaccharide in regulating the response to dietary antigens has been clearly shown in rodents, thus by the use of engineered strains of bacteria it may be possible to regulate the response to dietary proteins and so prevent the transient allergic responses that reduce the gut's ability to handle potential pathogens. This brief description of the regulation of the response to food proteins attempts to serve two purposes.

First, it may indicate a role for engineered strains in the controls of hypersensitivity reactions and the consequences thereof. Second, the type of response generated to food proteins differs markedly from that required to eliminate infectious agents. It is then perhaps not surprising, as will be discussed later, that dead mucosal microbial vaccines are generally not very effective.

THE RESPONSE TO INFECTION

The implications of biotechnology on animal health directly related to infections are numerous; it is therefore proposed to discuss them sequentially from diagnosis to vaccine development. The conventional methods for the identification of the causal agent have in the past depended heavily on an ability to isolate and grow the organism in pure culture. The use of monoclonal antibodies should aid both the identification of infectious agents and their virulence factors. Unlike conventional anti-serums, which by virtue of the nature of the immune response are very heterogenous, monoclonal proteins have a unique specificity which allows such discrimination to be made.

For example, by use of the enzyme-linked immunosorbent assay (ELISA) it is possible to detect subtle antigen differences between strains of an organism as well as to characterize the immune response of the infected animal. Once having identified a particular virus, for example as being important in the pathogenesis of diarrhoea, how then might one proceed towards an effective vaccine? Consider for instance that virus X has been shown to be an important pathogen but is closely related to similar commensals.

It can produce disease, and animals that recover from infection show solid protection. In neutralization tests it has been possible to show the specificity of X, but it is very difficult to grow! Further progress may then involve the development of a sensitive immunoassay for research and sero-epidemiology. Production of a monoclonal antibody (either conventionally from rodents or preferably from infected farm species) may allow isolation of a strain-specific antigen. Subsequently the antigen should be produced by cloning and expressing the gene in, for example E. *coli,* by recombinant-DNA technology. Such a procedure may form the basis for the development of a vaccine.

CONTROL OF INFECTION

Without attempting to consider ways of eradicating an organism from the environment there are two types of approach to the preventionxontrol of infection within an animal. The first is essentially passive and involves little or no requirement of the host's immune system, whilst the second requires the stimulation of an active vigourous immune response.

PASSIVE PROTECTION

Monoclonal antibodies have been used therapeutically to passively protect neo-natal pigs and calves against challenge with a lethal dose of a K99 positive pathogenic strain of E. *coli.* It has been pointed out that such an approach has a theoretical advantage over conventional vaccination of the dam, in that it does not require anticipation of the problem. Further, in this case where the disease is restricted to the first 24 h of life and the effects of the bacteria are contained within the intestine, it is directly treated by oral administration of the monoclonal antibody. A similar approach has been used to protect mice from intranasal infection with Respiratory Syncital Virus, but generally it would appear that the therapeutic use of monoclonal antibodies will be of benefit in specific, defined individual areas, notably in human medicine, rather than on a herd basis in veterinary medicine.

The rationale for this argument is the amount of antibody required, particularly if it is necessary that it is given orally. Monoclonal antibodies to infectious agents may, however, be of benefit in the 'education' of the immune system and this active process will be discussed later. Another passive form of treatment involves the use of anti-microbials, and in this context biotechnology has already made a significant impact. Human leucocyte interferon has been isolated by monoclonal anti-immunosorbents and on a commercial level recombinant interferon and interleukin 2 are now generally available.

ACTIVE PROTECTION

Before considering the potential impact of biotechnology on the development of safe, effective vaccines, it is necessary to consider what is required of such products in order that they may stimulate a protective response. As has already been mentioned, the vast majority of infectious agents are met at a mucosal surface and therefore vaccines should be directed towards stimulating protection at these sites. To this end it is clear that by far the most effective method of eliciting such a response involves presentation of the antigen at the mucosal surface. In general parenteral vaccines have not been effective, and where exceptions exist they almost always involve some mucosal component (*e.g.* oral + parenteral).

The other exception to this principle involves immunization of the dam to achieve passive protection of the sucking neo-nate, by virtue of the presence of antibodies in colostrum. In the case of domestic species, antibodies in colostrum are derived from serum and these may be effectively stimulated by parenteral immunization. Unlike colostrum, milk antibodies are locally produced and in this case, like other true mucosal secretions, mucosal immunization is required to stimulate production of antibodies.

Having established that mucosal vaccines are those of choice for both theoretical and practical reasons, it is, however, disappointing that it is surprisingly difficult to stimulate a vigourous, protective response via this route. As has already been pointed out, this is of benefit with regard to the 'harmless antigens' (*e.g.* dietary antigens) also presented to the mucosae. To date the majority of effective mucosal vaccines are living attenuated organisms reflecting the now generally held axiom that either replication or at least adherence are essential prerequisites of mucosal immunogenicity. Whilst there is now clear evidence of memory in the secretory immune system, vital if any vaccination regimen is to be effective, it has also been shown that the response is highly dose-dependent.

Thus where dead oral vaccines have been used, very large numbers of bacteria are required and even then the level of immunity is sub-optimal by comparison with that following the use of live organisms. Another problem associated with mucosal immunization, unique to the gastrointestinal tract, relates to an ability to pass through the stomach in a form that remains antigenic. In this context the use of capsules or liposomes may be of benefit. Alternatively the observed immunological link between mucosal sites, such that oral immunization leads to the appearance of antibodies in milk, may be of benefit.

The New Vaccines

Vaccines derived as products of biotechnology may take a variety of forms and rather than discussing them on a specific disease basis it is proposed to briefly describe the general feature of each type. The genes coding for protein or peptide antigens derived from a number of bacteria and viruses have been cloned and expressed in appropriate bacteria. For example, with the foot and mouth disease virus one of the structural capsid proteins, WI, has been identified as carrying the antigenic determinants necessary to raise neutralizing antibodies in experimental animals.

The viral RNA encoding for WI has been copied into double-stranded cDNA, inserted in a bacterial vector and expressed in *E. coli*. Subsequently a range of oligopeptides from VPI have been isolated or synthesized and a major antigenic determinant comprising nineteen amino acids identified. The question that remains is how these proteins or peptides may be presented to the immune system in order to stimulate a protective response. The isolated VPI protein, for instance, is not as antigenic as the virus and a 100 to 1000-fold excess on a molar basis is required in order to obtain satisfactory protection. With the oligopeptide it is necessary to couple to a carrier protein (*e.g.*, KLH) and although a single injection of 200 pg has been shown to protect guinea-pigs from virus challenge, to date no satisfactory method has been found to stimulate a protective response in cattle. Similar results have been obtained with *E. coli* antigens (K88ab, ac, ad and K99 have all been cloned and expressed) but it still remains to be determined how a protective mucosal response may be stimulated.

Whilst to date this synthetic peptide vaccine approach has been disappointing, its theoretical attraction over conventional dead vaccines remains since it avoids the requirement for handling large volumes of infectious materials, or vaccinal strains that grow poorly in bulk-suspension cultures, and the risk associated with failure to completely inactivate the virus before incorporation into a vaccine. An alternative approach involves the use of attenuated strains and these may take a variety of forms, and have been used against a number of agents. Deletion mutants of Aujeszky's disease virus have been prepared and although shown to have greatly reduced virulence are still highly immunogenic in pigs.

Similarly an attenuated strain of *Salmonella typhi,* Ty2la, which is unable to utilize galactose, has been derived. The organism is able to survive for a limited number of divisions, but antigenicity remains unaltered. For another enteric pathogen, cholera, an appropriate mutant is also available. In this case it produces only the B subunit of the toxin and thus, whilst being able to bind to the epithelial cell and give good protection, does not generally produce diarrhoea.

Where attenuated strains of a particular organism are not available or appropriate the use of live recombinant vaccine may be of benefit. Intranasal inoculation with recombinant *Vaccinia* virus has been shown to protect mice against upper and lower respiratory tract infections with influenza. Alternative vectors to *Vaccinia* are also available, for example yeasts or engineered non-pathogenic strains of E. *coli* that produce only the B subunit of the labile toxin, and so bind to enterocytes but do not produce diarrhoea.

The general acceptability of live attenuated vaccines remains to be determined and the possibility of developing techniques whereby peptide vaccines can be delivered in a way that stimulates a protective mucosal response remains attractive. In this context the use of peptide conjugates coupled to molecules such as cholera toxin or lectins that bind avidly to epithelial cells may be of benefit. Alternatively the potential role of antiidiotypic antibodies (monoclonal) may be of interest, it having been recently shown that such antibodies may prime for protection against E. *coli*. The capacity of biotechnology to produce a range of antigenic moieties derived from infectious agents would appear to be unlimited. Further, it should not be difficult to identify those antigens that are important in the pathogenesis of disease.

The major problem that remains is how best these products might be presented to the immune system in order to stimulate a protective immune response. It is therefore to this target of understanding how antigens are presented to and processed by the mucosal immune system that intense future research should be directed.

NEW HEALTH PRODUCTS FOR FARM ANIMALS

Animal diseases cost U.S. agriculture $17 billion each year, according to the Congress Office of Technology Assessment. To help prevent these losses, animal scientists are using biotechnology to develop an array of products to diagnose, treat, and prevent disease in farm animals. An area where animal biotechnology has already had a profound impact is diagnostic testing. Many animal diseases are difficult to diagnose. A veterinarian often has to wait hours or days for laboratory results to confirm a diagnosis. In the meantime, the veterinarian must either withhold treatment or risk using an inappropriate therapy. Using biotechnology, scientists are developing fast, accurate diagnostic tests for many of the most common farm animal diseases. In some cases, a veterinarian can conduct the diagnostic tests right on the farm and immediately start the best therapy.

Many of these new tests use monoclonal antibodies. Antibodies are proteins an animal's immune system produces in response to invasion by bacteria, viruses, or parasites. Each kind of antibody binds specifically with a characteristic part of a particular kind of invader. Scientists make monoclonal antibodies by fusing two kinds of cells. One is an immune system cell that produces antibodies, which bind to part of a particular disease-causing microbe; the other is a cancer cell. The cell that results from the fusion inherits from the immune system cell the ability to produce antibodies, and from the cancer cell the ability to reproduce indefinitely.

This new hybrid cell, or hybridoma, acts as an antibody factory. The antibodies are called monoclonal because they are produced by identical "clones" of the original hybridoma-all will bind with the identifying structure on the surface of the microbe in question. One application of monoclonal antibodies is as a test for brucellosis in cattle. This bacterial disease often causes cows to abort pregnancies, and it can be spread to farmers and people who drink milk from infected cows. Vaccines protect pregnant animals from abortion, but vaccinated cows may become carriers of the disease. Conventional diagnostic tests cannot distinguish between the disease-causing microbe and the vaccine, which is made from the microbe. Thus, these tests cannot help the farmer determine which animals are carriers of the disease and which are not. But diagnostic tests using monoclonal antibodies are so specific, they distinguish between cattle that carry the disease-causing bacterium and those that carry only the vaccine. With the aid of these new tests, farmers can isolate carriers to prevent the spread of infection. Diagnostic tests are not just for diseases.

A monoclonal antibody test to detect pregnancy in animals is so simple that a farmer can read the test right on the farm. Monoclonal antibodies are also used in a test to detect a disease called scours in piglets and calves. Scours is a bacterial disease of newborn animals that causes diarrhea. Many of the afflicted animals die of dehydration. By providing quick, accurate diagnosis, the monoclonal antibody test for scours permits immediate administration of therapy.

The therapy for scours in calves is also based on monoclonal antibodies. Infected calves are fed monoclonal antibodies that coat the offending bacteria, preventing the microbes from causing diarrhea. Some other therapies for animal diseases are based on recombinant DNA technology. Genetic engineering allows scientists to insert one or more new genes into an animal, plant, or microbial cell. In developing treatments for an animal disease, for instance, an animal gene is inserted into a microbe or an animal cell that can be grown in a laboratory culture. The inserted gene is one that produces a natural disease-fighting protein. The microbes or animal cells become minute factories, producing large quantities of the therapeutic protein. For example, genetic engineering is used to produce interferons and interleukin-2, natural proteins that fight viruses.

Veterinarians long have used antibiotics against bacterial diseases, but they have had few drugs to treat viral infections. Interferons and interleukin-2 not only kill viruses directly, they also stimulate the animal's immune system, improving the immune system's ability to fight disease naturally. One target of research regarding interferons and interleukin-2 is shipping fever, a disease of cattle caused by several viruses infecting an animal simultaneously. Although the animal's immune system normally keeps the viruses in check, outbreaks of the disease may result from stress during transport to feedlots or market. Shipping fever costs the beef industry more than $250 million a year. Injections of interferons or interleukin-2 before shipping may help to augment the cattle's natural immunity and prevent shipping fever.

BETTER VACCINES TO PREVENT DISEASE

In addition to producing new diagnostic tests and therapeutic proteins, animal scientists are using biotechnology to develop vaccines to prevent disease.

A safe, effective vaccine for swine pseudorabies, a fatal herpes-virus, is already in use. An advantage of the swine pseudorabies vaccine, and other vaccines made through recombinant DNA technology, is that they use only a small portion of the original microbe. Conventional vaccines use killed or weakened forms of the disease-causing microbe.

Sometimes, when the microbes are not completely killed or sufficiently weakened, a conventional vaccine may cause the disease it was supposed to prevent. Disease-causing genes are not included in the genetically engineered vaccines. Therefore, the recombinant vaccines build up the body's immunity without the risk of causing disease. Another advantage of recombinant vaccines is the speed with which they can be developed. "Conventional vaccine development can take 20 or 30 years, maybe even 100 years," says Karen Jacobsen, D.V.M., University of Georgia College of Veterinary Medicine.

"This is one reason why we don't have vaccines for a lot of animal diseases today". Recombinant vaccines are being developed for foot-and-mouth disease, a highly contagious viral disease that infects cattle, sheep, and other animals. Although this disease has been eradicated in North America, it still causes substantial livestock productivity losses elsewhere, particularly in developing countries. Infected herds must be slaughtered, and contaminated ranches must be left idle for months to prevent new outbreaks of the disease. Conventional foot-and-mouth vaccines are made by weakening the virus that causes the disease.

These vaccines sometimes revert to the virulent state, and they have caused outbreaks of the disease in Europe. The foot-and-mouth vaccine made with biotechnology cannot cause the disease because, as in other biotech vaccines, the disease-causing genes have been removed. The biotech vaccine does not need to be kept cold and, therefore, can be used in developing countries. In addition to being used in treating swine pseudorabies and foot-and-mouth disease, recombinant DNA technology is being used to develop vaccines against dozens of livestock and poultry diseases, such as scours and a variety of respiratory disorders.

Some vaccines may be injected into eggs so that chickens are immunized before they hatch. Most vaccines protect against viral or bacterial infections, but using genetic engineering, researchers in New Zealand and Australia have developed a vaccine against a parasitic disease. Farmers must destroy the meat from sheep infested with a species of tapeworm common in those two countries. The new vaccine, which protects sheep from the tapeworm, may save farmers millions of dollars annually.

PROTEINS FOR ANIMAL NUTRITION

Natural proteins called somatotropins, or growth hormones, help animals convert feed into muscle or milk. Using recombinant DNA technology, scientists have been able to develop bacteria that produce commercial quantities of somatotropins.

They have found that administering small additional amounts of these proteins to animals helps to increase the animals' efficiency of feed conversion, thereby saving the farmer money. Porcine somatotropin (PST) may help improve human health, as well as lower the farmer's cost of production. PST improves feed efficiency in hogs by 15 to 20 per cent. It also reduces fat deposition, allowing PST-treated hogs to provide consumers with leaner cuts of pork. The ability to produce lean pork has enormous implications for improving human health by reducing dietary fat and cholesterol. Another growth protein, bovine somatotropin (BST), is being administered to make dairy cows more efficient, thereby lowering the dairy farmer's costs. Regular administration of small amounts of BST increases a cow's milk yield by 10 to 25 per cent. Although feed intake also increases, this feed increase is proportionately less than the increase in milk production.

Studies indicate that BST can improve the milk-to-feed ratio by 5 to 15 per cent. Extensive laboratory testing has shown that the meat and milk of treated animals contain no more PST or BST than the meat or milk from untreated animals. Other tests have shown that these proteins do not act on the human body, and therefore, they are safe to eat. Nevertheless, the use of BST remains controversial. Testing is under way to make sure that animals are not adversely affected by treatment with somatotropins. Another protein that can be administered to animals to increase feed efficiency is called growth hormone releasing factor (GHRF). While not a growth protein itself, GHRF causes the animal to increase production of growth proteins.

IMPROVED ANIMAL BREEDS

Long before anyone knew about genes or DNA, people were selectively breeding animals to produce desirable traits, such as strength, feed efficiency, and disease resistance. Selective breeding is a crude technology, because each cross between two animals mixes hundreds of thousands of genes, both desirable and undesirable. Recombinant DNA technology is now making it possible for scientists to breed animals with great precision.

One or more new genes can be inserted into an animal embryo without disturbing the rest of the animal's genes. The resulting animal with the new gene or genes is referred to as a transgenic animal. Despite the differences in technique, the scientist using recombinant DNA has the same objectives as the selective breeder: improving efficiency, disease resistance, and other desirable traits. But recombinant DNA provides a wider array of possible new traits, since recombinant techniques allow animal scientists to insert genes not only from the same or closely related species, but also from distantly related animals, plants, and even microbes.

One possible application of gene transfer in animals is to develop dairy cows that produce more nutritious milk. Bottle feeding human infants with cow's milk is not an ideal substitute for breast feeding, because cow's milk differs from human milk in some important components. Scientists are looking for ways to transfer genes into dairy cows that will enable the cows to produce milk that more closely resembles human milk. Advances in Control of Parasitic Diseases Continue to Rely on Biotechnology Parasitic disease continues to be a major health problem worldwide in humans and domestic animals. Applying biotechnology has led to progress in prevention and control in many areas, particularly in regard to diagnosis.

Parasites, like most organisms, carry repetitive sequences in their genomes. These are sometimes tandem repeats (back-to-back in a chromosome), or they may be located at random throughout the genome. Often, repetitive sequences are intergenic and non-coding, and are thus far more subject to evolution and change, increasing the likelihood that they will be genus or even species specific. Since detection of these sequences is not dependent upon gene expression, which may vary over time, repetitive sequence-based assays allow detection of the parasite at any life stage.

The effectiveness of repetitive sequence-based polymerase chain reaction (PCR) methods of identification has been demonstrated for parasites of several clinically important genera (1). A PCR assay has also been developed to differentiate two closely related nematode species that are virtually indistinguishable by morphologic examination (2), as well as for diagnosis of Toxoplasma and Cryptosporidium infections (3). Attempts to develop anti-parasite vaccines have met with multiple failures. The problems often result from lack of post-translational modification by currently used vector-host systems and the inability to regenerate the secondary and tertiary structure of the native immunogens. Antigen presentation in recombinant parasite vaccines has also been thought to be a problem resulting in failed efficacy. There is now some evidence to indicate that cloning of genes for antigens of interest to yield chimeric molecules expressed in the BCG strain of Mycobacterium bovis may provide improved antigen presentation to stimulate both humoural and cellular responses (4); however, it is unlikely that future applications of this technology will be simple. Nonetheless, work is proceeding on the application of biotechnology to the development of vaccines against parasites in most food animals.

A particularly striking exception to the marginal successes reported in other systems has been a recombinant vaccine for a parasitic helminth of sheep (5). This product yielded greater than 94 per cent protection in experimental work and is now available as a commercial product. A further possibility is the application of anti-idiotypic monoclonal antibodies. When monoclonal antibodies are prepared against a monoclonal antibody, it is possible to obtain, in a very small number of cases, an antibody which mimics the original antigen. Results of work with Cryptosporidium perfringens epsilon toxin suggest that animals immunized with anti-idiotype monoclonals can be protected against toxin challenge.

IMPROVING HEALTH THROUGH DEVELOPING VACCINES

Most biotechnologies related to health focus on the needs of the developed world, meaning that 90 per cent of health research is devoted to the health problems of 10 per cent of the world's population. Two main approaches are being used to develop vaccines using recombinant DNA technology. The first involves deleting genes that determine the virulence of the pathogen, thus producing attenuated organisms (non-pathogens) that can be used as live vaccines. Currently, this strategy is more effective against viral and bacterial diseases than against parasites. Attenuated live vaccines have been developed against the herpes viruses that cause pseudorabies in pigs and infectious bovine rhinotracheitis in cattle. A number of candidate *Salmonella* vaccines have also been produced. The second approach is to identify protein subunits of pathogens that can stimulate immunity.

The International Livestock Research Institute (ILRI) used this approach to develop a vaccine against *Theileria parva*, the parasite that causes East Coast fever in African cattle. A novel strategy for developing vaccines against bloodsucking parasites involves using components of the gut wall of the parasite that are not usually exposed to the immune system of the host. When the parasite feeds, it ingests antibodies induced by the vaccine, which destroy the gut wall and, consequently, kill the parasite.

This strategy has been used successfully to develop a vaccine against the one-host tick *Boophilus microplus*. Vaccination is one of the most effective and sustainable methods of controlling disease. Vaccines against parasitic diseases in Africa and viral diseases in Asia have been shown to control disease effectively and increase livestock productivity. A recent approach has been to use vaccines based on DNA. The use of DNA in vaccines is based on the discovery that injecting genes in the form of plasmid DNA can stimulate an immune response to the respective gene products. This immune response is a result of the genes being taken up and expressed by cells in the animal after injection. The live-vector and DNA vaccination systems could be manipulated further to enhance the immunity conferred by the gene products. Experimental studies have demonstrated that these vaccines can potentially induce appropriate and enduring immune responses.

This technology is, in principle, one of the simplest and yet most versatile methods of inducing both humeral and cellular immune responses, as well as protecting against a variety of infectious agents. However, although immune responses have been induced in a number of larger species, most of the information on the efficacy of DNA immunisation comes from studies of mice. An exhaustive review of the information available on the use of DNA vaccines in farm animals, including cattle, pigs and poultry, has identified the areas that need specific attention before this technology can be used routinely.

These areas include the delivery, safety and compatibility of plasmids in multivalent vaccines and the potential for using immune stimulants as part of a DNA vaccine. Korean scientists have developed a combined vaccine against pleuropneumonia, pneumonic pasteurellosis and enzootic pneumonia in swine. Molecular biology has been used to produce an improved vaccine against swine fever. In the Philippines, a vaccine has been developed that protects cattle and water buffalo against haemorrhagic septicaemia, which is the leading cause of death in these animals. The new vaccine provides improved protection at a very low cost.

DIAGNOSTICS AND EPIDEMIOLOGY

Advanced diagnostic tests that use biotechnology enable the agents causing disease to be identified and the impact of disease control programmes to be monitored more precisely than was previously possible. Molecular epidemiology characterises pathogens (viruses, bacteria, parasites and fungi) by nucleotide sequencing, enabling their origins to be traced. This is particularly important for epidemic diseases, in which pinpointing the source of the infection can significantly improve disease control.

For example, the molecular analysis of rinderpest viruses has been vital in determining the lineages circulating in the world and instrumental in aiding the Global Rinderpest Eradication Programme. Enzyme-linked immunosorbent assays have become the standard means of diagnosing and monitoring many animal and fish diseases worldwide, and the PCR technique is especially useful in diagnosing livestock disease. Many diagnostic techniques currently used in developing countries are cumbersome and unsuitable for low-resource settings. Molecular diagnostic technologies that are either already in use or being tested in low-income regions include polymerase chain reaction (PCR), monoclonal antibodies and recombinant antigens.

These technologies can be modified to facilitate their application in the developing world. Simple hand-held devices that rely on the binding specificity of monoclonal antibodies or recombinant antigens to diagnose infection may be easily adapted for use in settings without running water, refrigeration or electricity. Molecular characterisation of the virus serotypes causing foot and mouth disease has helped in the vaccination and control programmes in Asia. In Japan and Taiwan, DNA testing is being used to diagnose hereditary weaknesses of livestock. One test looks for the presence of the gene responsible for porcine stress syndrome in pigs. Pigs with this gene tend to produce pale poor-quality meat because of their reaction to the stress of transport and slaughter.

Pigs with this gene can now be excluded from breeding programmes, so the gene will become less common. In addition, DNA testing is being used in Japan to check for the gene that causes leucocyte adhesion deficiency in Holstein cattle. Cattle with this condition suffer from gum disease, tooth loss and stunted growth. They usually die before they are one year old. By using DNA testing, carriers can be identified and eliminated from breeding herds. Bulls used for breeding can also be tested to make sure that they are not carriers. Another DNA test identifies a gene that leads to anaemia and retarded growth in Japanese Black cattle.

NUTRITION AND FEED UTILISATION

The shortage of feed in most developing countries and the increasing cost of feed ingredients mean that there is a need to improve feed utilisation. Aids to animal nutrition, such as enzymes, probiotics, single-cell proteins and antibiotics in feed, are already widely used in intensive production systems worldwide to improve the nutrient availability of feeds and the productivity of livestock. Gene-based technologies are being increasingly used to improve animal nutrition, either through modifying the feeds to make them more digestible or through modifying the digestive and metabolic systems of the animals to enable them to make better use of the available feeds. Feeds derived from GM plants (a quarter of which are now grown in developing countries), such as grain, silage and hay, have contributed to increases in growth rates and milk yield.

Genetically modified crops with improved amino acid profiles can be used to decrease nitrogen excretion in pigs and poultry. Increasing the levels of amino acids in grain means that the essential amino acid requirements of pigs and poultry can be met by diets that are lower in protein. Metabolic modifiers have also been used to increase production efficiency (weight gain or milk yield per feed unit), improve carcass composition (meat-fat ratio), increase milk yield and decrease animal fat. The use of recombinant bovine somatotropin (rBST) in dairy cows increases both milk yield and production efficiency and decreases animal fat.

In the USA, the use of rBST typically increases milk yield by 10 per cent to 15 per cent. Although trials conducted in developing countries have reported a similar percentage increase, this increase is not significant because of the low milk yields and the high cost-benefit ratio. However, rBST is being used commercially in 19 countries where the economic returns make its use worthwhile. A porcine somatotropin has been developed that increases muscle growth and reduces body-fat deposition, resulting in pigs that are leaner and of greater market value.

CONSTRAINTS ON APPLYING THE TECHNOLOGY

The application of new molecular biotechnologies and new breeding strategies to the livestock breeds used in smallholder production systems in developing countries is constrained by a number of factors. In the developing world, poverty, malnutrition, disease, poor hygiene and unemployment are widespread, and biotechnologies must be able to be applied in this context.

Over the last few decades, the green revolution has brought comparative prosperity to farmers with land, but the majority of farmers, who are landless or marginal farmers and subsist only on livestock, have been neglected and remain poor.

The major constraints on applying biotechnologies have been enumerated by Madan and include:

- The absence of an accurate and complete database on livestock and animal owners so that programmes can be implemented
- The biodiversity present within species and breeds in agro-ecological systems
- The fact that models of biotechnological intervention differ distinctly between developed and developing economies
- The fact that many animal species and breeds are unique to the developing world; each has its own distinct developmental, production, disease resistance and nutrient utilisation characteristics
- The lack of trained scientists, technicians and fieldworkers to develop and apply the technologies, both in the government and in the private sectors
- The absence of an interface between industry, universities and institutions, which is necessary to translate technologies into products
- The inability to access technologies from the developed world at an affordable price in order to make a rightful, positive and sustainable contribution to livestock production and the economic welfare of farmers
- The high cost of technological inputs such as materials, biologicals and equipment
- The failure to address issues of biosafety and to conduct risk analyses of new biologicals, gene products, transgenics and modified food items, and, above all
- The negligible investment in animal biotechnology.

The critical issues affecting livestock productivity have recently been re-examined. Research that aims to enhance productivity and sustainability should focus on improving livestock feeds and nutrition, improving animal health, managing natural resources relating to the livestock sector, assesing the impact of technological interventions, and strengthening the capacity of the national agricultural research systems of developing countries.

Furthermore, the potential production capacity and contribution of livestock to the economy are still not being achieved in developing countries because the transfer, adaptation and adoption of technology is hampered by the lack of a clear policy for livestock development that is conducive to the introduction of new proven technology and by the lack of information flow from and to decision makers. In developing countries, there is a wealth gap between urban and rural areas, which persists and may even be widening; the rural-urban divide also tends to be reflected in education and health indicators. In addition, women in rural (and urban) areas who are predominantly involved in animal husbandry have higher illiteracy rates than men. A survey of 21 African countries recently highlighted the substantial disparities in primary schooling between urban and rural areas, in favour of urban dwellers. Special attention must be given to the knowledge and information needed to enable rural people to apply biotechnology. There is a need to identify alternative delivery systems for animal health care and to propose new roles for the state and the private sector in service delivery.

BUILDING CAPACITY

Owing to the constraints outlined above, the economic benefits of animal biotechnology cannot be realised without a conscious, sustained, holistic, multi-stakeholder, participatory approach. There is a great need to ensure that capacity is not just created but also is retained and enhanced. Capacity-building activities must be carried out at all levels: the awareness of policy and decision makers must be raised, the necessary legal and regulatory frameworks must be initiated, the technical and regulatory capacities must be enhanced and institutions may need to be overhauled.

More importantly, it is necessary to assess and deploy competent operators and institutional capacity continuously so that, as biotechnology advances, the procedures required for its safe use can be constantly evaluated, upgraded and applied. This is a daunting task, but it can be achieved through firm commitment and partnerships.

FUNDING TO IMPLEMENT TECHNOLOGY

Developing and commercialising improved technologies in most developing countries has been the responsibility of the public sector, and technology has been disseminated freely. This situation will have to continue if superior genetics, diagnostics and vaccines are to be delivered. However, research and almost all commercial development of biotechnology in the developed world are being driven largely by the private sector.

The global trends in funding for research and development and production do not address the concerns, needs and opportunities of the developing world. Developing countries are finding it increasingly expensive to access and use new technologies.

There is limited private-and public-sector investment in animal health and production, particularly in relation to modern biotechnologies that are 'resource hungry'. Although several discoveries have been made in laboratories in the developing world, in most cases these have not been converted into useful technologies or products.

The key potential users–resource-poor often illiterate farmers with a limited knowledge base–do not feel that applying these technologies is worth the effort, cost and risk involved. This is mainly because there is no agency or industry that can scale up and package the technology. Also, in the developed world, there is an economic incentive to market biotechnological services and products; this is lacking in the developing world because of the limited purchasing power of resource-poor stakeholders.

Research in biotechnology in recent years has also been motivated by economic considerations, and little research is conducted in the developing world because of the probable lack of returns on the investment. For understandable reasons, current funding policies in developing countries focus on areas that will yield practical benefit in the short term.

In determining future policy, policy-makers and funding bodies must not lose sight of the substantial benefits that can be gained in the longer term by investing in strategic research into vaccine development. Adequate multi-institutional (national and international) support through an international donor consortium is needed to develop cost-effective, cheap and easily adaptable biotechnological products.

The amount spent by international agencies on animal biotechnology in developing countries is currently very low and constitutes only a small percentage of the total spending on agriculture. The World Bank, the Food and Agriculture Organization, the Consultative Group on International Agricultural Research, the United Nations Development Programme, the United States Agency for International Development, the Swedish International Development Cooperation Agency, the International Development Research Centre, the Asian Development Bank and other donor and funding agencies have to designate a higher percentage of funds to the livestock sector.

It has been convincingly shown that investing in livestock has a dramatic and far-reaching impact on the human development index. This is a strong argument in favour of investing heavily in animal production and health biotechnologies in order to bring economic prosperity, nutritional security, rural development and health improvements to poor populations in the developing world.

Although animal production is being changed significantly by advances made in thousands of biotechnology laboratories around the world, benefits are reaching the developing world in only a few areas of conservation, animal improvement, health care (including diagnosis and control of disease) and the augmentation of feed resources. Adopting biotechnology has resulted in distinct benefits in terms of animal improvement and economic returns to the farmers. Over the past decade, the ILRI has focused on biotechnological applications, especially in Africa, and several developing countries now have multi-institutional programmes to develop and apply biotechnology. The developing world will have to respond to the many genebased technologies now being developed with a sense of commitment, trained manpower, infrastructure and funding.

6

Animal Handling: Principles of Restraint and Handling

Calm animals will also provide more accurate research results that are less confounded by handling stress. Handling and restraint stresses can significantly alter physiological measurements. Beef cattle not accustomed to handling had significantly higher cortisol levels after restraint compared with dairy cattle that were accustomed to handling.

Prolonged 6-h restraint of sheep where they could not move resulted in extremely high cortisol levels of >110 ng/mL. Aggressive handling should never be used for farm animals. Multiple shocks with an electrical prod more than doubled the levels of lactate and glucose in pigs compared with careful handling without electric prods. Transportation performance standards include movement of animals with minimal risk of injury or death to animal or handler. Transportation is only performed when necessary. Making the transport experience more comfortable for each species should be a priority for animal handlers.

Training of animal care personnel in handling procedures should include consideration of the well-being of the animals. During the handling and restraint of animals, care should be exercised to prevent injury to animals or personnel. Animals should be handled quietly but firmly. Properly designed and maintained facilities operated by trained personnel greatly facilitate efficient movement of animals. Prolonged restraint of any animal must be avoided unless such restraint is essential to research or teaching objectives.

The following are important guidelines for the use of animal restraint equipment:

- Animals to be placed in restraint equipment ordinarily should be conditioned to such equipment before initiation of the project, unless the preconditioning itself would increase the stress to the animals.
- The period of restraint should be the minimum required to accomplish the research or teaching objectives.

- Electrical immobilization must not be used as a method of restraint. It is highly aversive to cattle and sheep. Electrical immobilization must not be confused with electrical stunning that causes instantaneous insensibility or electric prod use that does not immobilize animals.
- Restraint devices should not be considered normal methods of housing, although they may be required for specific research and teaching objectives.
- Attention should be paid to the possible development of lesions or illness associated with restraint, including contusions, knee or hock abrasions, decubital ulcers, dependent edema, and weight loss. Health care should be provided if these or other serious problems occur, and, if necessary, the animal should be removed either temporarily or permanently from the restraint device. Animals should be handled and restrained in facilities and by equipment appropriate for the species and procedure.

Some aggressive behaviours of larger farm animals pose a risk to the health and well-being of both herdmates and human handlers. These behaviours may be modified or their impact reduced by several acceptable restraint devices and practices. Only the minimum restraint necessary to control the animal and to ensure the safety of attendants should be used. Care should be exercised when mixing animals to minimize fighting, especially when animals are grouped together for the first time. Animals should be handled and restrained in facilities and by equipment appropriate for the species and procedure. For cattle, for example, a chute facility should be available. Unless they are very young or tame, calves restrained for routine procedures should be handled by means of a calf chute equipped with a calf cradle.

PRINCIPLES TO PREVENT BEHAVIOURAL AGITATION DURING RESTRAINT FOR ALL SPECIES

The following guidance is provided to prevent behavioural agitation:

- Non-slip flooring should be provided. Repeated small rapid slips may cause agitation.
- Avoid sudden jerky motion of either people or equipment. Smooth movements will keep animals calmer.
- When an animal is raised off the ground, during restraint, it will usually remain calmer if its body is fully supported.
- Even pressure over a wide area of the body has a calming effect. The Panepinto sling for small pigs and cattle squeeze chutes use this principle.
- A calm, confident tone of voice will help keep livestock calmer.

- *Optimum Pressure*: Not too loose and not too tight. An animal needs to be held tight enough to feel the feeling of restraint, but not so tight that it feels pain. Excessive pressure will cause struggling.
- *Blocking Vision:* Using a blindfold made from a completely opaque material will often keep cattle and horses with a large flight zone calmer. Solid sides on cattle chutes or a fully enclosed dark box have a calming effect.

EFFECTS OF VISUAL DISTRACTIONS AND HANDLING

Livestock of all species will often refuse to move through a chute or other handling facility if they see distractions such as shadows, reflections, or people ahead of them. Removing distractions that cause animals to balk and stop will facilitate animal movement.

A calm animal will stand and point its eyes and ears towards distractions that attract its attention. If the leader is allowed to stop and look at a distraction, it will often move forward and the other animals will follow. If the animals are rushed, they may turn back and refuse to move forward when they see a distraction.

Distractions are most likely to cause balking or other handling problems if the animals are not familiar with the facility. Experienced dairy cows will often ignore a distraction such as a floor drain, but new, inexperienced heifers will balk at it.

Table. Visual Distractions that may Cause Animals to Balk and Refuse to Move

- Sudden changes in floor structure or surface such as drain grates, objects on the floor or change in flooring material.
- Shadows, puddles, and shafts of light; seeing light through a slatted floor.
- Animals may refuse to enter a dark place. Use indirect lighting to facilitate movement towards the light. Animals tend to move from a darker place to a more brightly illuminated place, but they will not move into blinding light.
- Reflections on a wet floor or shiny metal. Move lights to eliminate the reflection or use non-reflective surfaces.
- Moving people in front of approaching animals. People should stand where approaching animals do not see them.
- Jiggling chains, coats on a fence, flapping plastic, or swinging ropes. Remove these distractions.
- Animals see people, moving objects such as vehicles or objects with high colour contrasts outside of the chute. Improve movement by installing solid sides.

Table contains a list of distractions that may cause animals to balk and refuse to move. This list can be used as a guide for modifying handling facilities where excessive use of electric prods is occurring. In facilities where animals move easily and quietly and electric prods are seldom used, removal of distractions may not be needed.

FACILITY DESIGN PRINCIPLES FOR ALL SPECIES

Flooring

For all species, non-slip flooring is essential. Animals often become agitated when they start slipping. Handling and restraint will be safer and animals will remain calm if animals have non-slip flooring. Handling facilities should have non-slip floors and good drainage.

Equipment Maintenance

Surfaces that contact the animals must be smooth and free of sharp edges that could injure animals. Sharp edges will cause bruises and injury. Managers should routinely inspect equipment and have a programme of regular maintenance based on use. Special attention should be paid to latches on restraint devices.

Sanitation

Managers should regularly inspect facilities to ensure cleanliness. When new facilities are being designed, ease of cleaning is an important part of the design. Concrete curbs can be used to direct manure to a drain. Hoses, shovels, and other tools that are needed for cleaning should be readily available. Sanitation equipment should be removed after routine cleaning.

Animal handling facilities should be regularly cleaned after use and maintained in good working condition. Injuries and accidents can happen to animals and handlers from equipment lockup or other problems that can occur with build-up of filth, breakage, or wear and tear. Managers should routinely inspect the facilities to ensure cleanliness and to maintain a regular maintenance schedule based on use.

ANIMAL PERCEPTION

HEARING

All species of grazing animals have sensitive hearing. Cattle and horses have hearing that is more sensitive compared with humans to high-pitched sounds. The human ear is most sensitive at 1000 to 3000 Hz and cattle are most sensitive to 8000 Hz. Handlers should not yell or shout at cattle because shouting may be just as aversive as an electric prod. In another experiment, the sounds of people yelling caused a greater increase in heart rate than the sounds of gates clanging.

Intermittent or high-pitched sounds caused greater behavioural reactions and increased heart rate in pigs compared with steady or low-pitched sounds. Intermittent sounds and rapid movements are also more likely to cause cattle to react. Handlers should be observant of the position of an animal's ears. Horses and cattle will point their ears directly towards things that attract their attention.

Vision

Cattle, sheep, and horses have wide-angle vision and they can see all around themselves without turning their heads. Grazing animals have depth perception when they are standing still with their heads down. Depth perception is probably poor when the animals are moving with their heads up. This explains why they stop and put their heads down when they see a shadow on the floor. Grazing animals are dichromats. The retinas of cattle, sheep, and goats are most sensitive to yellowish-green; and bluish-purple light.

The dichromatic vision of the horse is most sensitive at 428 and 539 nm. Dichromatic vision and the absence of a retina receptor for red may explain why livestock are so sensitive to sharp contrasts of light and dark such as shadows or shiny reflections on handling equipment. Poultry appear to have excellent vision. Chickens and turkeys possess 4 cone-cell types in the retina giving them tetrachromatic colour vision, compared with the human trichromatic vision based on 3 cone-cell types.

Moreover, the spectral sensitivity of chickens is greater than that of humans from 320 to 480 nm and 580 to 700 nm. Their maximum sensitivity is in a similar range to humans. The broader spectral sensitivity of poultry may make them perceive many light sources as being brighter than a human would see. Poultry may be more docile during handling in blue light spectra. Lighting conditions have a large effect on chicken behaviour when the birds are shackled for slaughter. During handling of poultry, the occurrence of flapping should be minimized. Changes in lighting may be used as one tool to keep birds calmer during handling.

AIDS FOR MOVING ANIMALS

Animals in properly designed facilities may be moved using their natural behaviour and without the use of any aids. The goals of movement should be to minimize stress to each individual animal, reduce fear, and maintain calmness in all animals. All handlers should be trained in the natural behaviour of the species including their flight zone and in proper handler movement and interaction, and be able to recognize any signs of distress, anxiety, or behaviours that may result in injury or stress to the animals.

When necessary, non-electrical driving aids such as paddles, flags, and panels may be an adjunct with the use of natural behaviour and handling skills. Handlers should be trained in the proper and effective use of each driving aid, which is appropriate to the species. An electric prod should only be picked up and used in a specific situation where it is needed and then put away. Handlers have a better attitude towards the animals when electric shocks are not used. Data collected at meat plants indicate that most cattle and pigs could be moved throughout an entire handling system without electric prods. On a ranch or feedlot, the use of electric prods should be limited to 10 per cent or less of the cattle.

When an electric prod needs to be used, it should be applied to the hindquarters of the animal. Usually 1 to 3 brief shocks are needed. If the animal does not respond, the use of the electric prod should be discontinued immediately. It should never be applied to sensitive areas of the animal such as the eyes, ears, genitals, udder, or anus. Battery-operated prods are recommended because they administer a localized shock between 2 prongs. Electric prods should not be used on newborn animals, debilitated weak animals, non-ambulatory downed animals, or emaciated animals.

Electric prods are highly stressful to pigs. Repeated shocks greatly increased the percentage of non-ambulatory pigs. Multiple shocks and aggressive handling significantly increased blood lactate and other indicators of metabolic stress compared with gentle handling. Pigs that become non-ambulatory because of fatigue or porcine stress syndrome should not have electric prods used on them.

Some examples of the use of an electric prod as a last resort or if human or animal safety is in jeopardy are listed below:

- To move an animal after repeated attempts with non-electrified driving aids such as a plastic bag on the end of a stick, flags, slappers, rattle paddles, or streamers tied to the end of a stick have failed; the use of an electric prod is preferable to beating, ragging, pushing, or hard tail twisting of animals. If excessive slapping or electric prodding is required routinely, then the personnel involved may be too anxious or inadequately trained in proper animal handling or the facility may need modifications. Smaller animals may be gently lifted or rolled onto a transport mechanism.
- To get a downed animal in a truck that is located at a truck stop on the side of a highway. In this situation, opening up the truck gates or unloading the animals is not possible. 3 For cattle that are choking in a head stanchion or headgate or become jammed in a chute or other equipment.

SAFETY AND ANIMAL HANDLING

Stockmen working in animal housing and associated handling facilities are exposed to a number of potential risks and health hazards. Many stockmen work alone, handling big, sometimes dangerous animals as well as large groups of animals.

Several work tasks involve daily lifting, carrying of heavy loads and awkward working postures, which may cause overload injuries on the worker's musculo-skeletal system. Other potential risk factors contributing to accidents as well as to physical health problems and physiological stress for the worker is the fact that they operate in an environment where they are exposed to airborne particles, gases, noise. The work environment is partly the same environment, which is provided for the animals, and it should be noted that there is a strong relationship between the animal welfare issues and the handling of the work environment problems in cattle houses.

HEALTH AND INJURIES AMONG LIVESTOCK FARMERS

Accident Agents

In an accident situation there is a complex interaction between man and injury agents. In animal husbandry the human is subjected to a high degree of exposure to a large number of factors that singly or in combination can be accident provoking, such as animals, equipment and building structures. Risky situations are often due to wet and slippery floors as well as systems for material and animal handling.

Prevention of Accidents

With reference to accident statistics some accident preventing facilities and measures can be listed:

- Good cattle handling and handling facilities
- Pens from which the herdsman can easily escape if necessary in case of for instance frightened cattle and aggressive bulls
- Personnel passages through yard fences, properly placed
- Wide feeding passage or possibility to shut out the cattle from the feeding area
- Non slip flooring
- Satisfactory lighting considering quality and quantity
- Well designed ladders and stairways
- Protective railings around the fodder hay-chute in two storey houses
- Handles and bolts that do not stand out in alleys and pathways
- Protectors for the knuckles on barrows
- Emergency stops on machinery

Table. Recommended Illumination Level for Different Work Areas in Buildings Used for Beef Production. Lux Values Refer to the Work Plane. In Addition to the Quantity, the Quality of Light must be Considered

Building or Work Area/Task	Intensity of Illumination, Lux
Animal housing area	150
Animal handling area	250

Veterinary treatment	1,000
Feed processing, feed alley	250
Silo, feed storage	100
Ladders and stairs	250

Ergonomics

Any normally fit adult person, regardless of sex, physical strength and body dimensions should be able to work comfortably in livestock buildings. Despite an increasing degree of mechanisation and automation many jobs in animal husbandry are still associated with lifting heavy loads, moving and carrying equipment, feed and other materials with manual methods.

An important step towards improved human welfare within cattle housing is to eliminate unnecessary loads on the human musculoskeletal system.

Health Aspects Related to Air Quality

It has been known for decades that farm workers suffer health risks due to inhalation of gases and airborne particles commonly referred to as dust. The particles may be liquid droplets or solid. Several respiratory disorders have been reported, which are associated with dust and work in confined livestock buildings such as hypersensitivity, pneumonia, acute inflammation, chronic bronchitis, occupational asthma and toxin fever.

The respiratory problems connected to gases range from mild irritations of the respiratory tract to lethal effects. For the most frequent components in the air environment in livestock buildings there are statutory threshold limit values, which differ from country to country. It should be noted that threshold limit values in connection with animal welfare in many countries often are lower than the work environment legislation.

Dust and Dust Reducing

Dust in animal houses is generated inside the building and arises primarily from the animals and bedding and feed material. The airborne particles are often carriers of biologically active material like endotoxins, pathogens and allergens that can highly be disease provoking.

Of special interest is the concentration (*i.e.* number of particles per air volume) of respirable particles with a diameter less than 5 μm that can be inhaled and deposited in the lung.

The smaller particle the higher possibility that the particle reaches the lower respiratory tract and the deeper the particle is deposited the higher is the risk for illness. Even if "common dust" can be seen by human eyes (larger diameter than 100 ìm), the very small respirable particles cannot be, so we are indeed dealing with a danger our senses are not able to detect.

There are basically four approaches that can be used for airborne dust reducing:

1. Prevent particle formation
2. Prevent particle release
3. Remove suspended particles from enclosed workspaces
4. Isolate workers from dust clouds in workspace.

Of course the best approach in this matter is to have the cattle and the working sites outdoors or in open barns. In the cattle barn the most effective way to obtain a low dust concentration level is to prevent formation by always using hygienically perfect forage and bedding material. Providing an effective ventilation system in a barn is a good practice to remove suspended respirable particles as well as gases.

Gases and Reduction of Gas Levels

In animal buildings over 150 different gases and volatile compounds have been identified. Most of the volatile compounds originate from biodegradation of animal excreta or are produced by the animals themselves such as carbon dioxide (by the respiration of the animal). The gases that are found in the highest elevated concen-trations are ammonia, carbon dioxide, hydrogen sulphide and methane.

It must be emphasised that it is dangerous to enter any manure tank without either using a self-contained air supply or combining several measures: testing the air, constant and adequate ventilation of fresh air and using an harness and lifeline on the person entering the tank. As well, human and animals positioned over a slatted floor with manure storage underneath are at risk during agitation, mixing or pumping of manure due to the gases, especially hydrogen sulphide. A good strategy is to evacuate dangerous zones during these manure-handling actions.

Ammonia

Ammonia is the most common polluting gas in the atmosphere of the cattle barn. Animal manure is the main source of ammonia. The gas together with hydrogen sulphide a major component in what is termed odour, noticeable at ammonia concentrations of 5 ppm or more. Higher concentrations than that cause irritation to the respiratory organs and can aggravate the negative health effects of high dust concentrations.

Many factors affect the emission and concentration of ammonia such as air flow rate, manure and air temperature, manure surface area, density of animals, the degree of mixing of urine and faeces, time intervals between manuring, the pH-value, carbon/nitrogen ratio, mois-ture content and type of bedding. Consequently, many measures can be taken such as: proper ventilation and sufficient number of air changes, low air temperature in the barn and good urine drainage characteristics of the floor.

Hydrogen Sulphide

Hydrogen sulphide is the most toxic gas in animal confinement houses and 200 ppm and more are lethal for humans. The presence of the gas is of concern in buildings with liquid manure systems particularly. During slurry clearing operations concentrations in the range of 1,500 to 2,000 ppm have been reported.

Since hydrogen sulphide is lethal the aim should be to keep it below detectable concentration at all working sites. Special care has to be taken when pumping or agitating the manure. During emptying a slurry cellar the workers and the animal should be evacuated from the barn. In mechanical ventilated, closed animal houses continuously evacuation of the gases in an indoor slurry pit through a perforated duct under the floor and along the pit can reduce the gas concentrations above the slatted floor. By removing the slurry daily from the cattle barn to an outdoor storage it is possible to hold the concentration at an acceptable level.

Methane

Methane is generated when manure is stored under anaerobic conditions. This occurs in all non-aerated manure storages. Primarily, methane is dangerous because of its flammability. If methane is mixed in a proper proportion with oxygen a spark will be able to set off an explosion.

Carbon Dioxide

Carbon dioxide is present in all air, indoor as well as outdoors. Only extremely high concentrations (70,000 ppm or 7 per cent and over) could have severe health effects.

Mostly, incidents and accidents involving carbon dioxide are connected with asphyxiation due to oxygen deficiency rather than the direct effect of carbon dioxide itself. For instance, it is dangerous to enter a haylage silo during storing. In animal buildings the gas is more an indicator of ventilation efficiency than a health risk.

Noise

Noise can be defined as "unwanted sound, rapid, annoying pressure vibration in the surrounding air". Ventilation fans, grinders, vacuum pumps and mechanical feeding and manure systems generate noise. High sound levels are a real stress factor.

Exposure to noise will be able to cause temporary deafness and permanent hearing loss after a variable period if the level is over 85 dB(A). The unit, dB(A), is exponential and the human sense of hearing will perceive the sound level as halving if a reduction of 10 dB(A) occurs. It is fairly easy to reduce the sound level of ventilation fans and other mechanical noise generators. Maximum noise level recommendations and regulations differ from country to country.

Furthermore, it is important to reduce noise when handling cattle in order to improve animal movement and to make cattle handling safer. Cattle are more sensitive to high frequency noise than human. Animals will be calmer and easier to handle, if noise level is reduced. Clanging and banging metal parts should be silenced with rubber pads. Equipment operated with hydraulics should be engineered for quietness.

CATTLE HANDLING AND CATTLE HANDLING FACILITIES

Handling facilities are an essential part of a safe, easy and rapid handling of cattle. Appropriate handling and handling facilities remove much of the stress and frustration of the workmen, which inevitably occurs with excited, stubborn or aggressive animals. Properly constructed facilities confine cattle safely and efficiently with minimal animal stress and risk of injury to both cattle and workers.

Animal Behaviour and Improving Animal Handling

Understanding cattle behaviour can help farm workers to avoid dangerous situations and minimise accidents to handlers, as well as to design the handling facilities appropriately. Animals have natural boundary called the Flight Zone.

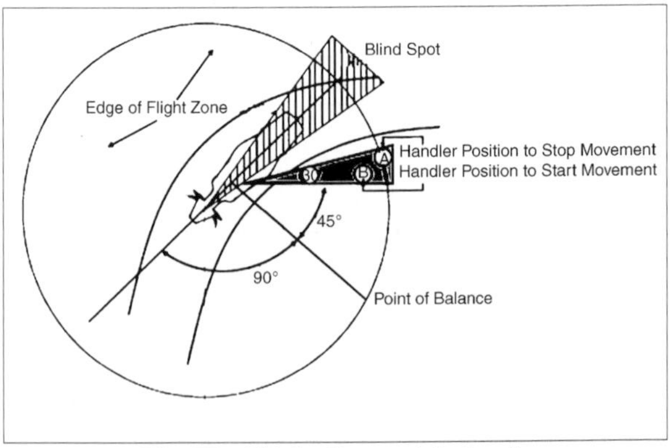

Fig. The Flight Zone of Animals

Deep penetration into the Flight Zone can cause panic and escape attempts. Handlers should have the possibility to remove, and should remove themselves from the Flight Zone, if the animal becomes aggressive. To move an animal forward: stand in the shaded area position B, behind the Point of Balance at the shoulder. Keep out of the Blind Spot at the rear of the animal. To stop movement: back off to position A. To make an animal back up: stand in front of the Point of Balance. To make an animal turn left or right: approach the animal head on. Factors reducing an animal's Flight Zone size: frequent contact with people; history of gentle handling; calm environment. Factors enlarging the Flight zone: infrequent contact with people; history of rough or abusive handling; excitement.

To improve animal handling and animal movement, many technical measures can be taken such as:

- Appropriate handling facility design.
- Animals tend to move from a dark to a more brightly lighted area. The light should illuminate the chute up ahead. Eliminate shadows and patches of light and dark, which may confuse animals. An approach is to illuminate the entire working area. Lamps should not shine into the eyes of approaching animals because glaring and blinding light impedes movement. Illumination should be uniform and diffuse.
- Prevent distractions, such as a chain hanging down in an entrance. Avoid sparkling reflection in a puddle, a moving reflection on a sheet of metal or bars of shadow across an otherwise sunlit alleyway. Dark colours can create shadow effects. Bright colours such as white and light yellow have been proven satisfactory.
- Reduce noise
- Animals might refuse to move, if they can see people ahead. Install shields to prevent animals from seeing farther ahead. Gates can be rigged with motor controls so a handler standing behind the cattle can open them.
- Solid sides that prevent the cattle from seeing outside the fences should be provided in the races and crowding pens. The crowding gate on the crowding pen should be solid as well, preventing animals from attempting to turn back to were they come from.
- Cattle are sensitive to changes in type and texture of floors and fences. Changes in type of flooring can cause balking. Use the same type of flooring throughout a facility, if possible. Use non-slip flooring. Drains should be located outside main drive alleys, chutes and crowd pens.

KINDS OF ANIMAL HANDLING

DAIRY CATTLE HANDLING

Mature milking dairy cows can be handled in head stanchions or a management rail. A complete squeeze chute is not required. Diagrams and pictures in Sheldon *et al.* illustrate methods for restraining tame dairy cows when they are held in a head stanchion. Young dairy heifers that are not accustomed to close contact with people are often handled most efficiently and safely in beeftype facilities with a squeeze chute. Disturbances by veterinarians and other visitors can reduce milk yield. If the cows are accustomed to many people walking through the milking parlor, there may be no effect because the frequent visitors have become part of their normal routine. Dairy animals are able to discriminate between people who have handled them in a negative manner and people who handled them in a positive manner. They were most likely to avoid the negative handler when he was seen in the same location where the aversive events occurred.

Dairy bulls are usually more dangerous than beef bulls. Bull attacks are a major cause of fatalities when people are working with livestock. One of the reasons beef bulls are safer is that they are reared in a social group on a cow. Price and Wallach found that beef bulls attacked more often when they were raised in individual pens. A dairy bull calf raised to maturity alone in a pen is more likely to be dangerous than a bull that was always kept with other animals.

If a bull is going to become dangerous, he is most likely to show aggression towards people at 18 to 24 mo. Handlers must learn to recognize signs of aggression that precede an attack such as the broadside threat. The bull will turn sideways to show how big he is before he attacks. Good descriptions are in Albright and Arave and Albright and Fulwider. Bulls that show aggressive tendencies towards people should be culled or transferred to a secure facility.

HORSE HANDLING

Teaching and research horses are usually handled using halters and lead ropes, and extra control may be achieved by using the chain of a lead shank placed over the horse's nose. Only trained horses should be tied and only to solid objects that will not give way if the horse pulls back. Lead ropes attached to the halter should be tied with quick release knot. Horses should never be tied with a chain looped across the top of the nose.

Cross-ties attached to each side of the halter should be equipped with panic-snaps or safety releases. A twitch may be applied to the horse's upper lip as a short-term restraint procedure. The movement of a horse may be restrained in stocks and chutes. An equine stock or chute may be as simple as a rectangular structure with a non-slip floor.

Other methods of restraint that may be applied by experienced individuals include front foot hobbles, sideline or breeding hobbles, or leg straps, but should be carefully considered depending on the training of the individual horse and the degree of restraint necessary. Chemical restraint can be effective and should be administered by a qualified person. With some drugs, an apparently sedated horse may react suddenly and forcefully to painful stimuli. General or local anesthesia should be administered by a qualified person, preferably a veterinarian, for painful procedures such as castration.

BEEF CATTLE HANDLING

Animals that are extensively raised and have large flight zones may become agitated if people stand close to the chutes and pens in the handling facility. If this occurs, solid fences may need to be installed so the animals do not see the people that are deep in their flight zone. Further information on facility design is in Grandin and Grandin and Deesing. There are many different designs of restraining chutes. Squeeze chutes should permit all animals to stand in a balanced position and the squeeze sides are applied evenly on both sides.

Squeeze chutes may be hydraulic or manual models. Settings of pressure relief valves for hydraulic restraint chutes should be adjusted to prevent excessive pressure from being applied. The chute should automatically stop squeezing at a reasonable pressure even if the operator continues to pull on the squeeze lever. A separate pressure control is required on chutes that have a hydraulic device for restraining the head. To avoid animal injury, this device must be set at a lighter pressure than other parts of the chute.

Pressure should be applied slowly to avoid exciting the animal. Excessive pressure can cause injury and incite cattle to fight the restraint. If cattle bellow the moment pressure is applied by a hydraulic device, this is an indicator of excessive pressure. Bellowing during restraint is associated with higher cortisol levels. Cattle should be able to breathe normally during restraint. The head gate can be self-catching or manually operated.

Self-catching head gates are generally not recommended for use with horned cattle unless they are operated manually. Unless they are very young or tame, calves restrained for routine procedures should be handled by means of a calf chute equipped with a calf cradle. Roping of cattle is necessary under certain conditions. However, roping should be performed by trained and experienced personnel and in a manner that minimizes stress to both the individual and the total herd. For head restraint of cattle in a squeeze chute, a properly fitted rope halter is recommended.

Nose tongs may be used on fractious animals in conjunction with other means of cattle restraint but nose tongs can slip and tear out of the nose, causing injury to both animal and personnel, and therefore are not recommended as a sole means of restraint. Nose tongs are aversive and cattle may resist the attachment of the tongs in the future. For repeated procedures that require head restraint, a rope halter is strongly recommended.

Electroimmobilization must not be used as a method of animal restraint; cattle and sheep find this procedure very aversive. Plastic streamers or a grocery bag tied to the end of a stick is an effective device for moving cattle and changing their direction. Cattle temperaments vary among individuals and among breeds. Handling should be adjusted for genetic and phenotypic differences.

SHEEP AND GOAT HANDLING

Sheep and goats show strong flocking behaviour in pens as well as on pasture. Breed, stocking rate, topography, vegetation, shelter, and distance to water may influence flocking behaviours. Isolation of individual sheep or goats usually brings about signs of anxiety. Separations from the flock, herd, or social companions are important factors that cause sheep and goats to try to escape. Sheep and goats tend to follow one another even in activities such as grazing, bedding down, reacting to obstacles, and feeding.

When handling sheep and goats, these characteristic behaviours should be considered and used advantageously and, more importantly, for the best interest of the animal's health and welfare. Transportation of sheep and goats should take into consideration the climatic conditions and productive stage of the animals. Care should be exercised in the transport of animals, and special consideration should be given during conditions of temperature extremes and high humidity.

Measures such as increasing the supply of nutrients immediately before long-distance transport that may reduce the risk of pregnancy toxemia and transport tetany in sheep and goats should be considered. Except for short distances when hauling is less physically taxing than trailing, transportation of ewes and does during late gestation should be avoided. When possible, animals should be gated off into smaller groups during transport to prevent pileups and death losses.

Additionally, temperature extremes or exposures should be considered and adequate and appropriate crating provided. Preventative or prophylactic medicinal agents may also be administered in an effort to minimize diseases that are associated with shipping. Sheep can be easily trained to enter a squeeze tilt table. The Panepinto sling can also be used for sheep.

Some restraint devices are more aversive than others. Welldesigned restrainers support the animal's body and do not have sharp pressure points. Both sheep and goats can be easily trained to enter head stanchions. Sheldon and Battaglia have illustrated guides on manual methods for holding sheep and goats. Designs for sheep races and corrals can be found in Barber and Freeman and American Sheep Industry Association.

POULTRY HANDLING

Poultry are handled in many experimental and teaching situations. Examples include wing-or leg-banding, immunization by intramuscular and subcutaneous injections, intranasal or intraocular application of drops and wing-web puncture, and removing or placing birds in different groups, cages, or holding and transportation crates. Injured, diseased or birds for transport should be euthanized on the farm. They should not be placed in transportation crates. People handling birds should be adequately trained so that stress to birds is minimal.

Poultry that are not familiarized to humans tend to struggle vigourously when caught. They can easily be injured if grasped improperly or subjected to excessive force. All poultry tend to flap their wings when caught, inverted, or caused to struggle for balance or footing. This tendency leads to risk of joint dislocation, bone fracture, or bruises when wings strike objects or other birds. The risk is particularly great for modern varieties of market-weight meat-type birds, which have powerful breast muscles but relatively weak joints due to their youth, or for caged light hybrid laying hens, which have fragile wing bones.

Poultry should be handled in ways that minimize wing-flapping or its harmful consequences. Care should be taken to prevent birds from striking their wings on door edges when placing them into or pulling them from cages or compartments. Particular care should be exercised in handling caged laying hens, which are prone to osteoporosis. To minimize the risk of bone fracture, hens should be held by both legs when removing them from the cage.

The manner in which a bird is carried can affect its fearfulness and stress. Broilers carried even briefly in the inverted position by the legs show a greater corticosterone response than do birds carried in an upright position, and the response lasts for about 3h. Therefore, birds should be carried upright whenever possible. Birds struggle less if they have been socialized, the body is fully supported in an upright position with wings restrained, the environment is relatively quiet, and the lighting is subdued.

Poultry should not be picked up or moved by one wing unless the wing is grasped near the base of the wing close to the body. They should quickly be released from such a hold, as when transferring birds from a coop to a floor pen. They should be shifted to a hold that firmly grasps both wings at their bases or that supports the body to minimize struggle and chance of a limb injury.

Ducks should not be caught by the leg because they are prone to leg injury if handled in this way. Large, strong birds such as turkey toms can be difficult to control by grasping a limb. They can also deliver punishing blows with their wings when struggling against capture. To pick up a very large turkey such as breeder tom, grasp one wing near the base of the body and then grasp the leg on the opposite side and set the bird's breast on the floor.

Finally, proceed with restraining the bird by grasping both legs. For intermediate-sized turkeys, the base of the wing and then both legs can be grasped simultaneously while lifting the turkey off the floor. Turkeys and ducks can be driven, so catching and handling of individual birds can be minimized by judicious use of alleys, ramps, and driving techniques when flocks must be relocated.

However, some birds such as older turkeys will not walk on different surfaces and therefore may have to be moved by individual handling. In many experimental and teaching situations, newly hatched birds or relatively small numbers of older birds need to be handled. In those cases, individuals can be easily caught and manipulated. Examples included wing-or leg-banding; immunization by intranasal or intra-ocular application of drops and wing-web puncture; and removing or placing birds in different groups, cages, and holding crates.

Trained and experienced scientists and caretakers know that birds struggle less if they have been socialized, if the environment is relatively quiet, and if the body is fully supported in an upright position. More complex procedures; for example, obtaining blood samples, intraperitoneal and venous puncture, and artificial insemination, often require at least 2 experienced persons. Skilled operators should adequately train personnel in such handling procedures so that stress to birds is minimal. Particular care should be exercised in handling caged layers to minimize the risk of bone fractures. When large numbers of birds housed under commercial conditions are to be moved or treated, handling methods need to be compatible with the housing systems involved.

A source of major concern should be the manner in which individual birds are caught, carried, and placed in new quarters or crates. In many situations, birds are at risk of injury because they are caught and moved by grasping a single wing with subsequent exertion of excessive force in moving the bird. No types of poultry should be picked up by one wing.

Gregory and Wilkins found that when laying hens were caught by one leg and removed from cages at the end of lay, the incidence of broken bones was 12.7 per cent; the incidence was only 4.6 per cent when both legs were used in removing hens from the cages. On commercial broiler farms, the chickens are usually picked up by a single leg. Leg breakage can be reduced if the birds are carried a short distance to the transport cage. When research is done under commercial broiler farm conditions, it is acceptable to pickup broiler chickens in this manner.

SWINE HANDLING

Snaring by the nose is a common method for holding swine for blood testing and other procedures. Good descriptions are in Battaglia and Sheldon. Snaring is probably stressful for pigs because they will attempt to avoid the snare after they have experienced snaring. For biomedical research, small pigs can be trained to enter the Panepinto sling. The animal is fully supported in a sling and its legs protrude out through leg holes.

A panel is the best device for moving pigs. Non-electric driving aids such as cattle paddles and flags can also be used by properly trained people. Guidelines on electric prod use are in the section on driving aids. Previous experiences with handling and the amount of contact with people will affect the ease of pig movement. Pigs with previous experiences of being calmly moved may be easier to move in the future. Calm, non-threatening movements of people will reduce stress levels in pigs and make them more willing to approach people.

7

Guidelines for the Care and Handling of Cattle

DAIRY CATTLE

Dairy cattle include replacement heifer calves and yearlings, dry cows, lactating cows, and breeding bulls used for research and teaching purposes related to milk production. The basic requirements for safeguarding the welfare of dairy cattle are an appropriate husbandry system that meets all essential needs of the animals, and high standards of handling. Physical accommodations for dairy cattle should provide a relatively dry area for the animals to lie down in and be comfortable and should be conducive to cows lying for as many hours of the day as they desire. Recent work indicates that blood flow to the udder, which is related to the level of milk production, is substantially higher when a cow is lying than when a cow is standing. Criteria for a satisfactory environment for dairy cattle include thermal comfort, physical comfort, disease control, and freedom from fear.

Cattle can thrive in almost any region of the world if they are given ample shelter from excessive wind, solar radiation, and precipitation. Milk production declines as air temperature exceeds 24°C or falls below–12°C for Holstein and Brown Swiss cows or below–1°C for Jerseys. Heat stress affects the comfort of cattle more than does cold stress. Milk production can be increased during hot weather by the use of sunshades, sprinklers, misters, and other methods of cooling as well as by dietary alterations.

Temperatures that are consistently higher than body temperature can cause heat prostration of lactating cows, but additional energy intake and greater heat production by the cow can compensate for lower temperatures, even extremely low ones. Consideration also needs to be given to humidity levels and wind chill factors in determining effective environmental temperatures. Adaptation to cold results in a thicker haircoat and more subcutaneous fat, which also reduces cold stress. Because dairy animals adapt well to cold climates, maintaining indoor air temperature equal to or slightly above outdoor air temperature is quite tolerable to housed animals. Coincidentally, providing the ventilation rate necessary to maintain this minimum temperature difference leads to good air quality.

Protecting the animal from extreme drafts, providing dry lying places that contribute to a dry, fluffy, erect haircoat, meeting the nutritional needs of the animal, and allowing the animal sufficient freedom of movement are essential. The newborn dairy calf has a lower critical temperature of 8 to 10°C. The intake of high-energy colostrum permits rapid adaptation to environmental temperatures as low as–23°C and as high as 35°C in dry, individual shelters with pens or in hutches.

Calves may be housed individually in outdoor hutches or inside buildings in bedded pens or elevated stalls. If calves are exposed to low temperatures, they should be provided with dry bedding and protected from drafts. Proper ventilation is critical in closed buildings with multiple animals. Hutches should be sanitized by cleaning, followed by moving the hutch to a different location or leaving the hutch vacant between calves.

In hot climates or during hot summer weather, calf hutches need to be environmentally modified or shaded to ensure that the calf does not experience severe heat stress. Housing and handling systems vary widely, depending on the particular use of the cattle in research and teaching. Recommended facilities for dairy cattle range from fenced pastures, corrals, and exercise yards with shelters to insulated and ventilated barns with special equipment to restrain, isolate, and treat the cattle.

Generally, headlocks, corrals, and sunshades are used in warm semi-arid regions. Pastures and shelters are common in warm humid areas. Naturally ventilated barns with free stalls are used widely in both warm and cold regions. To a lesser extent, insulated and ventilated barns with tie stalls are used in colder climates. Early research showed an economic advantage in providing housing for dairy cows during the winter instead of leaving them outside.

During good weather, to enrich the environment and to improve overall health and well-being, cows should be moved if possible from indoor stalls into the barnyard, where they can groom themselves and one another stretch, sun themselves, exhibit estrous behaviour, and exercise. Exercise decreases the incidence of leg problems, mastitis, bloat, and calving-related disorders.

Keeping cows out of mud and manure increases their productivity and reduces endoparasitic and foot infections. Current trends and recommendations favour keeping dairy cows on unpaved dirt lots in the southwestern United States and on concrete in the northern United States throughout their productive lifetimes.

Concrete floors should have a surface texture that provides good footing but does not cause injury. The concrete surface should be rough but not abrasive, and the microsurface should be smooth enough to avoid abrading the feet of cattle. Scraping a new concrete surface tends to remove microprojections formed during finishing Data are limited on the long-term effects of intensive production systems; however, concern has been expressed about the comfort, well-being, behaviour, reproduction, and udder, foot, and leg health of cows kept continuously on concrete.

As a safeguard, cows should be moved from concrete to dirt lots or pasture, at least during the dry period. An additional advantage is that the rate of detection and duration of estrus are higher for cows on recommended dirt lots or pastures than for cows on concrete. Exercise during the dry period does not adversely affect milk production, but does result in cows that are fit. Forced exercise after parturition reduces energy intake and milk production; therefore, forced exercise is not recommended.

Area

Between and within breeds, ages, and body conditions, critical dimensions of dairy cattle vary less with weight than with age. Body length and hip width are relatively uniform across breeds at weights between 180 and 450 kg. More than 94 per cent of the dairy cattle in the United States are Holsteins, and area recommendations for female calves and heifers are usually related to age groupings for Holsteins. Average normal growth curves relate heart girth and live weight to age.

The length of individual stalls should be a little longer than the length of the animal, defined as the distance between the pin bones and the front of the shoulders or between the pin bones and the brisket. For stanchions and tie stalls, stall width to length ratio should be at least 0.7. The width of free stalls should be twice the hip width. Dairy cows prefer larger, more comfortable stalls and use free stalls 9 to 14 h daily. Free-stall systems may be adapted for feeding trials utilizing electronic gates. Free stalls are recommended for dairy cattle used in teaching, extension, and research programmes throughout much of the United States.

Bedding

Resting dairy cattle should have a dry bed. Stalls ordinarily should have bedding to allow for cow comfort and to minimize exposure to dampness or fecal contamination. When handled properly, many fibrous and granular bedding materials may be used, including long or chopped straw, poor-quality hay, sand, sawdust, shavings, and rice hulls. Inorganic bedding materials provide an environment that is less conducive to the growth of mastitis pathogens. Sand bedding may also keep cows cooler than straw or sawdust. Regional climate differences and diversity of bedding options should be considered when bedding materials are being selected. Bedding should be absorbent, free of toxic chemicals or residues that could injure animals or humans, and of a type not readily eaten by the animals. Bedding rate should be sufficient to keep the animals dry between additions or changes. Any permanent stall surfaces, including rubber mats, should be cushioned with dry bedding.

Bedding material added on top of the base absorbs moisture and collects manure tracked into the stall, adds resiliency, makes the stall more comfortable, and reduces the potential for injuries. Bedding mattresses over hard stall bases such as concrete or well-compacted earth can provide a satisfactory cushion. A bedding mattress consists of bedding material compacted to 8 to 10 cm and enclosed in a fabric. Shredded rubber may be used and is recommended as mattress filler. Small amounts of bedding on top of the mattress keep the surface dry and the cows clean.

Ventilation

Ventilation permeates all aspects of the animal environment. Most often, ventilation is associated with respiratory health of animals: the quality of the air that animals breathe directly influences animal health and disease. Nevertheless, ventilation —directly and indirectly—affects many other aspects of animal health as well. Good ventilation in the lying area of lactating animals helps to keep bedding dry, a factor in favour of good mammary health.

Good ventilation along alleys helps to keep walking surfaces dry, a condition that contributes to healthy feet and a reduction in falling accidents. Good ventilation may lead to greater productivity; for example, maintaining air movement in the area of the eating area makes animals more comfortable, which is especially important during hot weather as an aid to maintaining dry matter intake. A comfortable, well-ventilated lying area encourages animals to lie down, an important contribution to many aspects of animal health.

During ventilation, outside air is brought into a barn where it collects moisture, heat, and other contaminants, all produced by the animals. Air is then exhausted to the outside. Ventilation is an air exchange process—contaminated air inside the barn is exchanged for fresh outside air. To determine ventilation rates, we focus on the moisture content of the air, as measured by relative humidity. But moisture is only one aspect; ventilation removes other undesirable contaminants as well. Ventilation is truly a process of dilution.

Air moved through a barn actually serves to dilute the inside air and, very importantly, to dilute all of its components. Dilution reduces concentrations of moisture and heat. Dilution also reduces concentrations of airborne disease organisms, harmful gases and dust, and undesirable odours.

The dilution rate of ventilation is often expressed in air changes per unit time. For example, a ventilation rate of 4 air changes per hour implies that the entire volume of the ventilated space is replaced every hour. In fact, some of the air may bypass the occupied zone in the barn, depending upon geometry of the space, the design of diffusers controlling inlet air, and so on. Thus, the effectiveness of ventilation is not often 1.0, but something less, perhaps 0.65. When ventilation is reduced below recommended levels—usually in a misguided effort to warm the barn using animal heat—less moisture is removed.

Sometimes the consequences of the resulting moisture buildup and lack of proper ventilation—usually condensation—are masked by:

1. Insulating the barn,
2. Using a greenhouse effect,
3. Providing supplemental heat, or
4. Dehumidifying the inside air.

For example, adding heat to the air reduces relative humidity, without the need for air exchange. It is quite possible to have substantial quantities of moisture added to the air and, if accompanied by heating of the air, have the relative humidity remain in an acceptable range. Thus, if relative humidity is the only measure of air quality, air quality may be deemed satisfactory.

However, even though excess moisture may not be apparent, the reduced dilution does indeed result in increased concentrations of airborne disease organisms, harmful gases and dust, and undesirable odours. If these increases are ignored, animal health problems are inevitable. Underventilation in winter is one of the most serious threats to the environment of animals.

Improper design and improper management of the ventilation may be reasons that wintertime ventilation is lacking, compromising animal health. Problems are most likely during winter, spring, and fall, especially during rainy weather and warmer days coupled with cold nights. Specific recommendations for ventilation system design are available. In general, minimum ventilation is provided by a continuous rate in winter, amounting to at least 4 to 6 air changes per hour. Summer ventilation rates may range up to and above 90 air changes per hour. Maintaining good air quality is a fundamental aspect of that healthy environment with ventilation providing the key. Through ventilation, the air inside the barn is continually diluted, ensuring that the air the animal breathes has low concentrations of all contaminants that threaten the animal's health.

TYPES OF HOUSING

In colder climates, stanchion and tie-stall barns have served well for herds ranging up to 50 or 60 milk cows. However, stall barns are labour intensive, both for milking and feeding. Comfort or tie stalls are preferred over stanchions. To avoid contamination of the teat and reproductive tract orifices, manure removal must be more regular and thorough when cows are housed in tie stalls. Cow trainers and gutter grates are recommended to ensure cleaner stalls and cows. Free-stall barns are a type of loose housing with one free stall recommended for each lactating cow. Depending upon provisions for feeding, different groups of cows can be fed differently according to their particular nutritional requirements. This has led to barn arrangements that permit division of milking herds into groups, usually by production.

One free stall is recommended for each lactating cow. The stall base and bedding provide a resilient bed for cow comfort and a clean, dry surface to reduce the incidence of mastitis. Because cows prefer to stand uphill, the stall base should be sloped forward 3 to 4 per cent from rear to front. Commonly used materials for the base include concrete, clay, sand, and stone dust; hardwood planks tend to rot. Rubber tires, if not firmly imbedded, tend to become loose.

In an ideal free stall, the stall bed and partition should define the lying position of the cow and accommodate natural lying and rising behaviour. Proper free-stall care includes daily inspection and removal of wet bedding and manure, in addition to adding dry bedding periodically. Neglected free stalls with excessive moisture or accumulations of manure can lead to an increased incidence of mastitis. For stalls with bases that must be replenished such as sand, an upward slope of the base towards the front should be maintained. This upward slope helps position cows more squarely in the stall when lying down, which contributes to cleaner stalls and cleaner cows. Free-stall hardware and other components should be kept in good repair.

Corrals should be scraped as needed and concrete alleys should be scraped or flushed regularly to clean them effectively. Feedbunk areas should be scraped regularly and any leftover feed removed. Shades and corrals should be designed to minimize areas of moisture and mud. Pastures must be managed to avoid disease transmission. Stocking rates should maximize production per head unless forage supplementation is provided or unless production per unit of pasture area is to be studied.

This strategy minimizes the stress that may result from overgrazing and minimizes ingestion of plants from areas immediately surrounding those areas contaminated with excreta, thereby reducing the challenge of potential pathogens and helminth parasites. Some pathogenic microbes may survive more than 6 mo in fecal deposits. Shade should be provided during hot weather.

Special Needs Areas

Cows with special needs are associated with greater risk and thus require special consideration with respect to facilities:

- *Preparturition*: Cows that are near the time of calving benefit from a clean, dry environment and access to an appropriate dirt lot for exercise. Feeding facili-ties should be provided to prepare cows for the high-energy ration they will receive upon entering the milking herd. Free-stall housing situated for frequent observation and proximity to the maternity area is a desirable option.

- *Maternity*: In preparation for calving, cows should be moved to individual pens that are separate from other animals, especially younger calves. The environment should be well ventilated, and the pens should be maintained to be clean, dry, and well bedded. Recommended pen size is 3.7 m × 3.7 m or 3 m × 4.3 m. The maternity pen should have a stanchion on one side for cow restraint. A concrete curb between each stall aids sanitation. Deep bedding should be used on concrete floors to prevent cows from slipping. Grooved concrete is also recommended. Provisions should exist for lifting downer cows. Devices to aid and promote standing include hip lifters, slings, inflatable bags, and warm water flotation systems. Pen location should permit access by a tractor or loader to allow removal of downed cows. All downed cows should be promptly examined by a veterinarian and handled in a humane and appropriate manner. Each pen should be provided with adequate feeding space and fresh, clean water. Depending on local conditions, a calving pen may not be necessary. Cows can calve in a pasture area with lighting situated for observation. A calving pasture should be well sodded and drained, should be large enough to allow cows to move away from others in the group before calving, and should contain an adequate sheltered area. Use of a pasture pen can eliminate footing and bedding problems associated with calving pens.
- *Removing Calf*: Dairy calves are normally removed from their dams as soon as possible following birth. The cow and calf are more difficult to separate after 3 d. Therefore, early removal is recommended. To prevent transmission of Johne's disease, follow the National Johne's Education Initiative control programme.
- *Postcalving*: A cow that has recently calved should be placed in a special area for frequent observation before rejoining the milking herd. Individual feed intake and milk production should be monitored to determine whether the cow is progressing normally. Milk must be withheld from shipment as required by regulations. Free stalls or large, well-bedded pens may be used in this special area. For a larger herd, a special hospital and maternity barn, possibly equipped with a pipeline or portable milker, could house cows in this management category as well as cows that are calving or that have other special needs.

- *Treatment*: A treatment area in the barn is recommended for confining cows for artificial insemination, pregnancy diagnosis, postpartum examination, sick cow examination, surgery, and for holding sick or injured animals until recovery.
- *Dry-off*: Cows recently dried off should be separated from the milking herd for feeding purposes. Recommended medical treatments should be performed, and cows should be observed frequently to ensure normal progress.

Lighting

Lighting recommendations for dairy cattle housed in indoor environments are the same as those for beef cattle in intensive environments.

FEED AND WATER

Except as necessary for a particular research or teaching protocol, dairy cattle should be fed diets that have been formulated to meet their needs for maintenance, growth, production, and reproduction. Feed ingredients and finished feeds should be wholesome, carefully mixed, and stored and delivered to the cattle to minimize contamination or spoilage of feeds. To ensure freshness, feeds that are not consumed should be removed daily from feeders and mangers, especially highmoisture feeds such as silage. Feed should be far enough from waterers to minimize wetting of feed. Space should be adequate for feed and water. Feeders or mangers should be designed with smooth surfaces for easy cleaning and increased feed consumption. The recommended linear space per cow at the feed bunk is 61 to 90 cm, which should allow every animal uninterrupted feeding. Feeder design should permit a natural head-down grazing posture to promote intake, improve digestive function, facilitate normal tooth wear, and decrease feed-wasting behaviour.

At least one water space or 61 cm of tank perimeter should be provided for every 15 to 20 cows in a group. At least 2 watering locations should be provided for each group of cows. Each cow in tie stalls and stanchions should have its own water bowl or drinking cup. All calves should consume colostrum in amounts of 8 to 10 per cent of body weight within 4 to 5 h after birth always before milk is fed, and another 2 to 3 L within 24 h of birth for a 36-to 45-kg calf.

Colostrum should be monitored with a colostrometer for quality. Mixed highquality colostrum pooled from several cows can be better than low-quality colostrum from a particular dam. However, it is currently suggested that individual cows be tested for colostrum quality for the use of their colostrum alone with avoidance of colostrum from known disease-carrying cows. Proper handling and storage of the colostrum is essential.

Until calves can consume dry feed at an adequate rate, they should be fed liquid feed in amounts sufficient to provide needed nutrients at a rate up to 20 per cent of body weight at birth per day until weaned. Water should be given at times other than when milk or milk replacer is fed to avoid possible interference with curd formation. However, this is not a problem with most milk replacers currently fed. Fresh water should be provided at all times. Replenishment of water should follow milk or milk feeding by at least 15 minutes. Calves being raised as replacement heifers or for beef should be fed enough dry feed with sufficient fibre preweaning to stimulate normal rumen development. Calf research guidelines have been reported that permit uniformity in measuring and reporting experimental data.

Water intake affects consumption of dry matter and is itself influenced by individual behaviour, breed, production rate, type and amount of feed consumed, water temperature, environmental temperature, atmospheric vapour pressure, water quality, and physical facility arrangement. Non-lactating cows consume 3 to 15 kg of water/kg of dry matter consumed, depending on environmental temperature.

Lactating cows consume 2 to 3 kg of water/kg of milk produced plus that required for maintenance. Water should be available at all times; it should be checked daily for cleanliness and monitored regularly to ensure that it is free of contaminants that could potentially put zoonotic agents into the human food chain. Water sources should be readily accessible to all stock. Underfoot surroundings in watering areas should be dry and firm. Cattle should not be able to wade in drinking water.

SOCIAL ENVIRONMENT OF HUSBANDRY

Dairy cattle are social animals that exist within a herd structure and follow a leader. Cows exhibit wide differences in temperament, and their behaviour is determined by inheritance, physiology, prior experience, and training. Cattle under duress may bellow, butt, or kick; however, cows are normally quiet and thrive on gentle treatment by handlers. Cows learn to discriminate among people and react positively to pleasant handling.

Aversive handling leads to more incidents during handling and transport for calves than positive handling. Similarly, heifers and cows exposed to aversive handling took longer to traverse and more force to move than those handled more gently. Although the presence of an aversive handler reduced kicking during udder preparation, residual milk was 70 per cent greater than for the control milkings.

Cows have higher milk yields if handlers touch, talk to, and interact with them frequently. Cows should have visual contact with one another and with animal care personnel. Handling procedures are more stressful for isolated cattle; therefore, attempts should be made to have several cows together during medical treatment, artificial insemination, or when cows are being moved from one group to another.

This was verified by increased heart rate, hypothalamic-pituitary-adrenocortical axis activity, and vocalizations. Pain sensitivity is reduced during isolation, suggesting a stress-induced analgesia. Care should be taken to minimize the negative impact of moving cows to new groups by avoiding frequent regrouping and by always moving more than one animal at a time to a new group. The use of a trainer cow can have a positive impact on adjustment to feedlot environments where many heifers are raised. However, dairy calves had few indicators that repeated regrouping and relocations stressed calves. Aggression was rare and the calves seemed to habituate to the repeated mixing.

Calves from larger groups after weaning had fewer incidences of displacement of other calves from the feed barrier, were more active, and had more positive interactions with familiar calves. Calves, like cows, prefer familiar calves to unfamiliar calves during stressful situations, and a familiar companion calf improved cows' reaction to separation.

Social status can affect health issues such as lameness. Low-ranking cows spent more time standing and standing half in cubicles than did middleand high-ranking cows. Standing half in cubicles correlated positively with the number of soft tissue lesions related to lameness. Dairy cattle have traditionally been kept in groups of 40 to 100 cows, although specific research protocols may require smaller or larger group sizes.

Variation in group size—small, medium and large—does not cause a problem *per se*. Expansion to a larger herd size, however, can affect management decisions because overcrowding with an insufficient number of headlocks or inadequate manger space per cow, irregular or infrequent feeding, and excessive walking distance to and from the milking parlor have a greater impact on behaviour and well-being than does group size. Cattle of all ages are gregarious.

Socially isolated cattle show clear signs of stress: increased heart rate, vocalization, defecation/urination, and cortisol levels. In addition, there are benefits of housing cattle together. For example, pairs of calves are more likely to play than isolated calves, a behaviour thought to be associated with positive welfare. Young calves should be kept in groups from 2 to 7 animals in order for animals to benefit from social contact, but larger groupings are associated with health problems and morbidity.

Management of resources is an important part of reducing aggression and other problems, such as cross sucking, in groups of animals. Adult dairy cattle should have 1 freestall/cow to reduce competition. Similarly, dairy cattle with more space at the feedbunk engage in fewer aggressive interactions and the reduction of competitive behaviour associated with more feeder space is particularly marked in post-and-rail feeder design.

Cross sucking in calves is an undesirable behaviour performed in groups. Calves are typically fed 10 per cent body weight during the milk-fed period and there is clear evidence that this feeding level is insufficient. A combination of slower milk flow, hay feeding, and access to a non-nutritive artificial teat are also recommended to reduce cross sucking. Providing additional objects for oral manipulation, such as tires, has also been shown to reduce other problems such as stereotypic tongue rolling in calves.

Restraint and Handling

Vaccination schedules that are appropriate for the location and dynamics of the individual herd should be established with the advice of the attending veterinarian. Certain dairy cattle behaviours put at risk the health and well-being of herdmates as well as the humans handling the cattle. These behaviours can be reduced or modified by implementing principles of low-stress handling and restraint that include appropriate movement of people, well-designed facilities, optimal lighting, non-slip flooring, and smooth, quiet restraint devices.

Stanchions, head gates, and squeeze chutes can be modified to function optimally, but acclimation and positive reinforcement by individuals trained in low-stress handling can minimize the need for additional restraint by halters, rope, tail hold, and nose tongs. Hobbles and casting ropes should be used selectively and only when necessary. Chemical sedation is always preferable to excessive use of force or application of electrical prods. Information about calving management is given by Albright and Grandin.

First-calf heifers should be bred to calving-ease bulls and be of appropriate stature and body condition to minimize the chances of dystocia or the need for calving assistance. Optimal calving conditions in a clean, quiet environment with employees appropriately trained to follow calving protocols will result in more live calves and fewer calving injuries and illness.

Calving injuries should be assessed immediately so that appropriate footing is provided and proper treatment is implemented. Cows that are unable to stand should be moved to a soft-bedded pack and examined by a veterinarian within 2 to 4 h of calving. Calves require special handling and care from the time they are born. Colostrum should be fed or ingested within the first 5 h after birth always before milk is fed. Between 1.89 L and 3.79 L of colostrum are necessary to impart adequate immunity to the calf.

In the absence of colostrum, a colostrum replacement product that delivers at least 125 g of immunoglobulin should be given by bottle, bucket, or tube feeder. Colostrum is rich in nutrients and provides the calf with vital immunoglobulins and other important immune factors. Clean navels can be dipped in a dilute chlorhexidine solution as soon as possible after birth. Good nutrition as supplied by a combination of milk starter grain, and fresh water along with proper handling and close monitoring starts a calf on its way towards a healthy life.

STANDARD AGRICULTURAL PRACTICES

All animals should be individually identified. Heifer calves should have supernumerary teats removed at an early age. Removal may be performed in the first 3 mo of life with a scalpel or sharp scissors. Older calves and heifers close to calving that have supernumerary teats should be examined by a qualified person. The removal of extra teats at this advanced age is necessary if they will later disrupt the milking process or be at risk of becoming infected. If so, they can be removed with proper restraint and use of appropriate anesthesia by a qualified and trained person. Milking procedures should follow National Mastitis Council guidelines. Routine breeding programmes should include housing and handling facilities that allow for effective implementation of artificial insemination programmes. Castration may be performed on male calves.

Foot Care

Lameness in dairy cattle is a major source of economic loss to the farmer and a serious cause of pain and discomfort to the cow. It is perhaps the most important condition affecting the welfare of cows on dairy farms. Lame cows suffer lowered milk production and reduced fertility, and are culled at 2 to 4 times the rate of healthy control cows.

The pain associated with lameness results in changes in the animal's gait that include:

- Arching of the back;
- Shortening of the stride length on the affected limb;
- Sinking of the dew claws on the unaffected contralateral limb;
- Head bob in a vertical plane;
- Reduction in walking speed, and frequent stops; and
- Swinging the affected limb in or out depending on the location of the painful lesion.

These alterations can be used to provide a locomotion score for each animal, and the most commonly used system in North America utilizes a 5-point system of scoring where 1 is non-lame and 5 is severely lame. Herd workers should be taught how to score locomotion so that they can identify cows with scores >2 for treatment by an attending veterinarian or hoof-trimmer. Around 85 per cent of lameness in dairy cattle is associated with lesions in the rear feet, particularly the outer claw, because of the overgrowth of horn resulting from the redistribution of weight as the cow walks on hard concrete surfaces, with a large udder occupying the space between her rear legs. This overgrowth of the outer claw may be removed and the weight transferred equally between the inner and outer claw by regular hoof-trimming.

Trimming to restore a normal toe length along the dorsal hoof wall of around 75 mm for mature Holstein cattle, combined with balancing weight between the inner and outer claw, lasts around 4 mo on average. Therefore, it is recommended that cattle be trimmed at 6-mo intervals, typically at the time of dry off and in mid-lactation around 90 to 150 d in milk. Some cows with pre-existing hoof disease may require attention more frequently.

Hoof lesions causing lameness may be broadly classified into 2 groups:

1. Infectious and
2. Claw horn.

Infectious lesions include digital dermatitis, interdigital phlegmon, and heel horn erosion. These lesions are associated with poor feet and leg hygiene and are a particular problem in free-stall environments, where the cow is exposed to alleyways contaminated with wet manure when she is not occupying a stall. Putative agents such as several species of *Treponema* and *Fusobacterium necrophorum* are involved in the pathogenesis of these conditions, but hydropic maceration of the skin of the interdigital space appears to be a prerequisite for the development of disease.

Infectious causes of lameness are controlled by improving leg hygiene by removing manure from the walkways and by the use of a topical antibacterial administered either directly to the lesion by a hand-held spray or via a footbath. The frequency of foot bathing is dependent on the degree of manure contamination of the cows, and a variety of chemicals are available for use, such as copper sulfate, zinc sulfate, and formalin.

Use of any of these chemicals should be done under veterinary direction. Claw horn lesions include sole hemorrhage, sole ulcer, toe and heel ulcer, and white line disease. These are clinical signs on the surface of the claw that represent the result of several possible causative pathways. Sinking of the third phalanx within the claw horn capsule, due to a breakdown in the connective tissue of the suspensory apparatus, may be caused by hormonal changes at calving time and nutritional events such as subacute ruminal acidosis. Sinking of the third phalanx compresses the corium below, interrupting the flow of blood and nutrients to the cells responsible for horn growth. As a result, a defect develops that becomes apparent several months later as the sole horn continues to grow.

Excessive removal of sole horn, either through poor hoof trimming or due to excessive wear from walking long distances on rough concrete will also contribute to lesion development. Flooring surfaces should be non-slip, avoid excessive trauma to the claw surface and be dry. Concrete should be grooved to improve traction; a pattern that utilizes parallel grooves 3/4 inch wide and deep, spaced 3 inches on center appears to provide a good compromise between sufficient traction to reduce injury while limiting the amount of wear.

For transfer lanes between milking centers and the living accommodation, a 1-m wide strip of rubber flooring has been used successfully to reduce trauma and wear, and rubber flooring has been used in parlor holding areas to provide cushion for cows that have to stand for long periods of time. The severity of the claw horn lesions that develop is influenced by the time spent standing each day, which results in increased loading of the claw and increased compression of the tissues below the third phalanx.

Time spent standing may be increased by:

- Poor stall designs that fail to provide surface cushion, room to lunge, and sufficient resting area;
- *Overstocking*: Providing fewer usable stalls than there are cows in a pen;
- Excessively prolonged milking times;
- Time spent locked up away from the stalls for management tasks; and
- *Heat Stress:* Cows may stand more in an attempt to cool off.

In addition, lame cows struggle to use stalls with hard surfaces because the act of rising and lying down becomes more challenging due to foot pain. These cows stand more in the stall and fail to gain adequate rest for lesion healing. For this reason, deep sand-bedded stalls provide the gold standard in cow comfort. If sand stalls are unavailable, lame cows should be treated and returned to a bedded pack area for rest and recuperation until normal ambulation returns.

Failure to identify a claw horn lesion early in its course may result in deep digital sepsis. This is a complication caused by infection of the deeper structures of the claw, including the distal interphalangeal joint and tendon sheaths. Such animals are usually severely lame and require euthanasia or extensive surgery. Seeking veterinary assistance is recommended for individual cows that show signs of lameness or if a significant lameness issue exists for the herd.

Disbudding and Dehorning

A review of horn anatomy and growth and dehorning and disbudding of cattle was provided by the AVMA. The AVMA also provides guidance on use of sedation, anesthesia, and analgesia and alternatives to horn removal. The AVMA policy on dehorning/disbudding should be followed. Calves should be observed closely for 1 to 2 h following dehorning. No food or water should be offered until the sedation is completely worn off or reversed. Persistence of a depressed attitude, head-pressing, or an abnormal head tilt for more than 2 h should result in a complete examination.

Tail Docking

The bovine tail has several physiological and behavioural functions including dissipation of heat, and facilitation of visual communication among cattle and with human caretakers; the tail often serves as a primary mechanism of fly control. Removal of the lower portion of a cow's tail is commonly referred to as "tail docking" and the use of tail docking as a routine dairy farm management tool apparently originated in New Zealand.

New Zealand farmers responding to a 1999 survey believed removal of tails resulted in faster milking, reduced risks to the operator, and reduced rates of mastitis. Similar unsubstantiated claims have been made for the US dairy industry. Several European countries, some Australian states, and California have prohibited tail docking. Both the Canadian and American veterinary medical associations have policy statements that oppose the practice of tail docking for routine management of dairy cattle. The policy statement of the *American Association of Bovine Practitioners* (AABP) indicates that scientific evidence to support tail docking is lacking and recommends that "if it is deemed necessary for proper care and management of production animals in certain conditions, veterinarians should counsel clients on proper procedures, benefits, and risks.

Scientific studies have been performed to evaluate both the potentially negative and positive aspects of tail docking. Important welfare issues that have been evaluated have included pain caused by tail docking, changes in fly avoidance behaviour, immune responses, and changes in levels of circulating plasma cortisol. Experiments that have been performed on both calves and preparturient heifers have consistently concluded that the process of tail docking does not induce significant acute or chronic changes in plasma cortisol or other selected physiological measures. Modest changes in general behaviour of calves that have been docked using rubber rings or cautery irons have been reported but these changes have not been associated with significant differences in normal feeding, ruminating, or grooming behaviours. Likewise, few significant differences in general behaviour of docked preparturient heifers have been noted.

However, greater changes have been observed in tail surface temperatures of docked heifers compared with heifers with intact tails, indicating that heifers may experience chronic pain similar to the phantom pain reported by human amputees. Research has demonstrated that tail-docked heifers flick their tails more often and are forced to use alternative behaviours such as rear leg stomps, feed tossing, and head turning to try to rid themselves of flies. More flies settle on tail-docked cows than on intact cows, and the proportion of flies settling on the rear of the cow increases as tail length decreases. In another study, there were no significant differences in the numbers of stable flies found on the front legs of cows but docked cows had nearly twice as many flies on their rear legs compared with those with intact tails.

Fly avoidance behaviours were increased in the docked animals, whereas tail swinging was increased in the control animals. Foot stamping was identified only in docked animals and, overall, fly numbers and fly avoidance behaviours were increased in docked animals. Researchers have been unable to identify improvements in udder health or udder cleanliness for animals in commercial herds that have docked tails.

In one study, the effect of tail docking on cow cleanliness and *somatic cell counts* (SCC) was evaluated over an 8-wk period for lactating cows that were housed in a free-stall facility. Standardized cleanliness scores obtained from the rump, midline of the back, or rear udder were not significantly different between docked and intact animals nor was there any significant difference in SCC or the number of teats containing obvious debris. In another study, SCC, occurrence of *intramammary infections* (IMI), and udder and leg hygiene scores were evaluated over an 8-mo period for lactating dairy cows that had been blocked by farm and randomly allocated to tail-docked or control groups.

No significant differences were found in SCC or udder and leg hygiene scores. The prevalence of contagious, environmental, and minor pathogens was not significantly different between cows with docked or intact tails. Although current studies do not indicate that the process of tail docking modifies physiological indicators of stress, several studies have documented changes in fly avoidance behaviour and recent research has suggested that docked tails have enhanced sensitivity to heat.

No benefits to cattle welfare have been associated with tail docking. The routine use of tail docking in research or teaching herds should be discouraged, and alternatives to tail docking are recommended when appropriate. Any use of tail docking, other than for medical reasons, should be reviewed and approved by the IACUC.

SPECIAL CONSIDERATIONS OF MILKING MACHINE AND UDDER SANITATION

The milking facility should have a programme for regular maintenance of milking machines and follow the recommended mastitis control programme of the National Mastitis Council. Appropriate equipment and competent personnel should be available for milking. Personnel responsible for milking should receive ongoing training about proper milking procedures as the frequency of training has been associated with adequacy of milking performance.

Animal care facilities should be designed and operated to standards meeting or exceeding those of grade A dairies as defined in the Pasteurized Milk Ordinance. Areas where milking takes place must be designed and constructed in accordance with the 3-A Sanitary Standards Inc. Accepted Practices. Cows should be maintained in housing areas that provide for adequate hygiene to ensure that udders are visibly clean. Cows should be milked on a regular schedule that is appropriate for the goals of the herd or specific research project.

Written operating procedures should be established to control potential contamination of milk with antibiotics or other pharmaceutical agents. Antimicrobial treatments should be administered based on approved defined protocols. All extra-label treatments must be administered under the supervision of a veterinarian that has an appropriate veterinary-client-parent relationship. Milking machine and udder sanitation are vital to an effective preventive programme against mastitis and follow guidelines as established by the NMC.

Care should be used to minimize the excessive use of water before and during udder preparation. Emphasis should be placed on ensuring that cows enter the milking parlor with clean, dry teats. Udders, especially teat ends, should be clean and dry when teat cups are applied for milking. The removal of foremilk before teat disinfection is encouraged as a means to detect mild cases of clinical mastitis. Teat sanitation, predipping, and wiping immediately before machine attachment reduce udder infection caused by environmental pathogens. Postmilking disinfection of teats is an essential management practice that greatly reduces the incidence of mastitis. Milkers handling cows should pay meticulous attention to their own personal hygiene and wash their hands thoroughly before milking and frequently during milking.

The use of clean nitrile or latex gloves during milking is highly encouraged to prevent contamination of the udder. Cows with subclinical cases of contagious mastitis should be milked last to reduce the spread of mastitis throughout the herd. Udder hair removal is recommended as a means to improve milking hygiene and udder health. Cleaning of milk handling equipment is accomplished by a combination of chemical, thermal and physical processes and cleaning regimens should be designed to meet appropriate regulatory standards.

Recommended cleaning and sanitizing practices are a balance between the cleaning temperatures, cleaning chemical concentration, contact time and mechanical action. Effective cleaning programmes for milking machines include use of hot water; use of disinfectant solutions and other chemical agents effective for removing mineral, milk fat, and protein deposits from equipment between milkings; disinfection of teat cups between cows; and flushing of teat cups with warm water, cold water, boiling water, or chemical disinfectant solution.

The most common routine in the United States is a combination of prerinse, alkaline detergent, acid rinse and premilking sanitize. Very small herds may utilize manual cleaning and disinfecting that involves hand-cleaning of some or all of the milk harvesting and storage equipment. Small to medium herds commonly use automatic washing equipment. This equipment will automatically mix the chemicals with the appropriate water volume and temperature and circulate these solutions through the milking machine. On large farms, an attendant may be present to mix chemical solutions and operate valves for circulation.

The effectiveness of milking system cleaning can be evaluated by examination of standard plate counts and laboratory pasteurized counts performed on bulk tank milk samples.

Stray Voltage

The term stray voltage describes a special case of voltage that develops on grounded metal objects on farms. If this voltage reaches sufficient levels, animals coming into contact with grounded devices may receive a mild electrical shock that can cause a behavioural response. At voltage levels that are just perceptible to the animal, behaviours indicative of perception such as flinches may result, with little change in normal routines.

Studies by numerous independent research groups in several countries are in agreement that the most sensitive cows begin to react to 50 or 60 Hz electrical current of 2 mA applied from muzzle to hooves or from hoof to hoof. This corresponds to a contact voltage level of about 1 V. As the voltage and current is increased, a greater percentage of cows will react with behavioural responses becoming more pronounced. Numerous studies have documented avoidance behaviours at levels above the first reaction threshold. The median avoidance threshold for 50 or 60 Hz current flowing through a cow is about 8 mA. Even when the threshold is exceeded not all cows would be expected to show a behavioural response but as the voltage increases, signs in a herd would be expected to be more widespread and uniform. The scientific evidence strongly suggests there is no relationship between behavioural responses to stray voltage and physiological or hormonal responses.

There is no apparent relationship among behavioural modifications, milk production, and animal health. The only studies that have documented adverse effects of voltage and current on cows had both sufficient current applied to cause aversion and forced exposures. It is typical for voltage levels to vary considerably at different locations on a farm. Decreased water and feed intake or undesired behaviours will result only if current levels are sufficient to produce aversion at locations that are critical to daily animal activity. These locations include feeders, waterers, and milking areas. Controlled research has shown that if an aversive voltage was administered to a water bowl once per second, water intake was reduced. However, when the same voltage was applied once every 10 min and once per day, no reduction in water intake was observed. If an aversive current occurs only a few times per day, it is not likely to have an adverse effect on cow behaviour. The more often an aversive voltage occurs in areas critical to cows' normal feeding, drinking, or resting, the more likely it is to affect cows.

No one sign is pathognomonic; a variety of signs has been reported in cows exposed to different levels of voltage. Documented signs are behavioural changes and decreased drinks of water per day and length of time per drink. The amount of water consumed may not be affected even when behavioural modification occurs. Intermittent periods of poor performance, poor milk letdown, and incomplete or uneven milk-out, abnormal behaviour during milking, increased milking time, refusal of feed or water, increased SCC in milk, and increased mastitis are signs often attributed by farmers to stray voltage; however, none of these signs were evident in numerous controlled studies. These signs are often caused by other factors such as abusive cow handling, faulty milking machine, poor milking techniques and hygiene, and nutritional deficiencies. Therefore, animal behaviour or other symptoms cannot be used to diagnose stray voltage problems. The only way to determine if stray voltage is a potential cause of abnormal behaviours or poor performance is to perform electrical testing. A thorough investigation of the entire production unit should be conducted to determine other sources of problems. Electrical systems should comply with wiring codes and standards at all times to protect both animals and people.

Whenever suggestive signs cannot be attributed to other causes, measurements should be taken to determine if a voltage potential exists, and the results recorded for future comparisons. A diagnostic confirmation of stray voltage must include a competent electrical measurement indicating at least 2 to 4 V between 2 points that a cow might contact, with some cows should exhibiting avoidance behaviours at this location. Voltage levels may need to be monitored at different times of the day and on different days because the threshold level may be exceeded intermittently.

All voltage readings should be made with a 500 to 1,000 Ω resistor across the 2 measuring leads to the cow contact points in addition to open circuit measurements. Readings without the use of a shunt resistor are meaningless. Although the resistance of cow and human tissues is similar, the contact resistance is generally lower for cows than for humans, particularly if cows are in a wet environment. The resistance of a cow's body plus the contact resistance with the floor is commonly estimated as 500 Ω. This is a reasonable value for a cow standing on a wet floor.

Cows standing on a dry surface will typically produce 1,000 Ω resistance or higher. Cows standing or lying on dry bedding will have a resistance many times higher than this. The resistance of a human can be as low as 1,000 Ω for wet hand-foot contact to >10,000 Ω for dry handfoot contact. The contact voltage to produce sensation can therefore be higher for humans than for cows, depending on the conditions of the contact points. If more than 1 V is detected at the cow contact points, it is advisable to have a qualified electrician or the local power supplier evaluate the situation.

Bulls

The feeding and watering of growing and mature bulls should meet requirements of the National Research Council. Bulls should be housed in clean, well-lit, and ventilated buildings or outside in facilities that protect them from inclement conditions and allow them to remain clean and dry. Young bulls kept in small and uniform groups should be observed carefully as they mature to make certain that one or more individuals are not injured. A panel can be installed in the center of group-housing pens to allow subordinate bulls to escape aggressive behaviour of dominant pen mates. Aggressive behaviour increases with age, and group housing should be discontinued by around 3 yr of age. Smaller or subordinate bulls should be removed from the group, and a bull removed from a group for over a few hours should never be returned to the group.

Visual and vocal social interactions with other bulls may be stressful. The safety of humans and animals is the chief concern underlying bull management practices. By virtue of their size and disposition, bulls may be considered as one of the most dangerous domestic animals. Management procedures should be designed to protect human safety and to provide for bull welfare. Electroejaculation of bulls is sometimes necessary and should be performed by a qualified person using equipment that functions properly and is in good repair.

A programme of annual self-regulation should be followed for:

- Semen identification and sire health auditing service and
- Minimum requirements for health of bulls producing semen for artificial insemination.

COOLING PONDS FOR DAIRY CATTLE

Cattle have used water for cooling probably for as long as there have been cattle. The last survey done in Florida reported that 30 per cent of the dairies had cooling ponds. The reader should be aware that cooling ponds, if not maintained properly, could harbour infectious diseases such as Leptospirosis and many mastitis organisms. The water usage is very high to fill and keep full.

Environmental concerns that must be addressed are the run-off from ponds and ground seepage. Every state seems to have many regulations that must be addressed before building cooling ponds. The benefits of cooling ponds and some suggestions on building and maintaining them. It is the dairymen's responsibility to use them properly and to be aware that ponds can harbour diseases and the pond water must be dealt with according to state and local laws.

What is a cooling pond for dairy cattle? For this discussion they are a man made hole in the ground with a constant inflow of clean fresh water. Usually the ponds drain or overflow from the top. If a series of ponds is used, the overflow water may be run into a retention pond to be irrigated on crops or enter the waste management system. Cooling ponds may also be fed from springs or artesian wells. So cooling ponds for dairy cattle should have constant fresh water in and a constant outflow of water to some place for proper disposal.

Cement ponds may also be built. These may drain from the bottom and can be used to flush feed lanes or barns. Bottom draining will help remove some of the solids that have settled to the bottom of the pond. Care must be taken on the entrance and exit to cement ponds. These may be very slippery if the cement is not cross-grooved. A stagnant pond without fresh water entering, a lagoon, a wallow, and a mud hole do not classify as a cooling pond for dairy cattle.

Site Selection

A site near the feed and drinking water, ideally with some form of shade for cows, is best. The idea is for cows to cool, leave the pond and goes eat and drink, lay down in the shade and then do it all over again. If cows do not have feed and drinking water near, they will stay in the pond and not eat and we have cool wrinkled cows that do not give much milk.

Cooling ponds should be located close to a water supply as fresh water needs to enter the ponds constantly. There must be a way to dispose of run-off water properly. This will vary by areas of the country. Some slope from the ponds to the waste management area would be helpful, so pumps need not to be used. Cooling ponds should not be too far from the milking parlor as long walks in the sun increase heat stress. Exit lane sprinklers help make the long walk more comfortable.

Concrete Ponds

Concrete ponds will work and have been used but they are expensive to build. However, they are easier to maintain, as the entrances and exits do not have to be rebuilt. The finish of the concrete on exit and entrance slopes should be cross-grooved to prevent cows from falling. A 1:8 slope would be a safe slope if concrete ponds are constructed.

Sizing Cooling Ponds

There is no official listing for cooling pond size. If a holding area requires 15/ft^2/ cow and cows are driven into a pond and taken out as a group then 15/ft^2/ cow might be adequate. If 50/ft^2/ cow is recommended for most general animal space, then this might be the correct figure for a cooling pond. This amount of space would allow cows to enter and exit the pond at all times when all cows were not using the pond at the same time. If 50/ft^2/ cow for a pond is used, then 5000/ft^2 is then needed for the 100 cow group.

Pond Shapes and Depths

Using the sample of 5000/ft^2/100 cows a pond could be 50' wide × 100' long, or a circle 80' in diameter (A= (Pi) r^2; A= 3.14 × 40 = 5024ft^2). If a 50' × 100' pond is used and the sides are fenced so the cows can only enter or exit at each end of the pond, the amount of dirt falling into the pond will be reduced. If a circular pond is made, cow carpet should be used to eliminate the problem of dirt falling in the pond.

As cows can swim, pond depth is not a problem. Deeper ponds may allow the settling of organic matter on the bottom of the pond. This organic matter may not be disturbed by cows walking through it as they will swim over the top of it. Preference should be given to depths of 3' or 4' as it seems to cool the cows and let them stand and walk in and out at will. While deep ponds allow cows to submerge, they do not seem to like to float too long so they go to where they can stand and may block cow traffic. A circular pond that is deep in the middle lets cows dunk themselves while others enter and exit the pond at will.

Water Usage

A 100 cow group with 50/ft^2/ cow could have a 50' wide × 100' long × 3' deep cooling pond (it may be 4' in the middle and slope in and out). A cooling pond 50' wide × 100' long × an average 3' deep=15000/ft^3. There are 7.48 gallons/ft^3, thus the pond requires 112,200 gallons of water.

The pond must be filled and then kept fresh with running water at all times while in use. A 1" pipe 300 ft. long and water pressure at 35psi will result in a flow rate of about 23.5 gpm. At this rate it would then take 80 hours to fill the pond initially. The dairy water usage would then be 1,410 gph × 24 hours or 33,840 gallons of water per day. This fresh water must be pumped into the pond and there is going to be close to that amount of run off water that will exit the pond for disposal.

Water Run-off

Disposal of run-off water must be in accordance with whatever local, state, and federal laws that applies. Many dairies run the water into the waste management system, while others add it to retention ponds to be used for irrigation.

Shading over Ponds

Experience has led to discontinued use of shade over ponds as the cows never leave the ponds. Remember, the object of this exercise is to increase dry matter intake and cows will not eat much soaking in the pond.

Pond Maintenance

At least once a year, usually during the winter, the pond should be fenced and the water removed by pumping. The mud and manure on the bottom of the pond should be removed and the pond left empty for the sun to shine on it. In the spring, the pond should be filled again and the fences removed. Some people just fence the pond in the winter, dig a new pond and fill in the old one with new pond's dirt.

Mycoplasma mastitis is often cultured in water from cooling ponds. This is a real problem if pulsators do not work well resulting in damaged teat ends. Many herds with mycoplasma positive ponds, have no mycoplasma in the bulk tank. If the water is filthy, cows should not be permitted in the ponds and the ponds should be pumped dry. The sludge should be cleaned and the ponds filled with water again. Both the sludge and the water shoud be removed from cow areas.

Cooling Ponds to Reduce Body Temperature

In the summer of 1986, an experiment was conducted at a North Florida dairy that was using man-made cooling ponds. Ten early lactation Holstein cows were fitted with radio transmitters that transmitted inner ear temperatures every five minutes to a radio receiver and data logger. Data were collected over a four-day period in mid-August. The cows' activities were also monitored during this time. Entering and exiting the ponds, eating, drinking and laying were recorded.

The cooling ponds lowered the cows' temperature by 1-2°F depending on the time of day they entered the cooling ponds. The average length per stay in the pond was 18 minutes for events from midnight to noon and 12 minutes per visit from noon to midnight. It was obvious that cooling ponds cooled cows.

Pond Water Quality

In 1987, samples were taken from a dairy's man-made earthen ponds. The ponds were sampled weekly from May 19,1987 to September 28, 1987. Six to ten ponds were sampled each week.

Total bacteria counts (TBC), varied greatly from week to week. This may have been due to no water entering the ponds. There did not seem to be any great increase in total bacteria counts or coliform counts as time progressed. The total bacteria count averaged 3,133,700 CFU/ml for all ponds for the 20 weeks. The coliform counts averaged 14,340 CFU/ml for the same period.

Cooling Ponds and Their Effect on Milk Quality

In a previous study on a dairy in Florida with man-made cooling ponds, it was reported that cows exposed to cooling ponds did not experience more clinical mastitis than cows that did not have access to cooling ponds. In fact, cows exposed to cooling ponds during the trial period were only half as likely to develop a case of clinical mastitis as cows not exposed to cooling ponds.

Clinical Mastitis Organisms

The same dairy had provided its clinical mastitis records for several years. The total number of clinical mastitis cases was high in the first quarter of the year and then declined sharply about the time hot weather came and the cows started using the ponds. The incidence stayed low, even in the last quarter of the year when the ponds were not in use. This can be explained many ways. The first quarter of the year was an extremely wet period. The cows were quite dirty and the cow wash system was available to one-half of the herd. This may explain some of the variation, even though one-half of the herd had never had a cow wash. Another confounding variable was that this herd switched to Clorox for pre- post-teat dipping in April of that year. It had previously pre- and post-dipped with a Chlorhexidine product. From this it could be concluded that Clorox caused the reduction of clinical mastitis. However, there were no controls in this study, one could conclude nothing except that the ponds did not increase clinical mastitis.

This herd was on a lactation treatment study during the summer of 1987. A total of 40 cows were sampled and treated. There were no unusual organisms treated. These 40 cows were not all the cows treated for clinical mastitis during the period the cows had access to the ponds, but the results were encouraging.

Statewide Effects of Ponds on Milk Quality

Regulatory samples for all the herds in Florida were obtained from the Division of Dairy Industry for the year of 1987 and were analysed for the effects of no ponds, man-made ponds and natural ponds. The numbers presented are least squares estimates of Standard Plate Counts (SPC) and Direct Microscope Somatic Cell Counts (DMSCC) for each month. The data are not biased by different numbers of dairies with no ponds, natural, or man-made ponds. The data were analysed by herds whose milking cows had access to ponds as indicated on the survey.

The data indicated that milking cows with access to man-made cooling ponds had lower SPC and DMSCC counts than other herds in the state with no access to ponds and those herds with natural ponds were higher in both categories. From these data it could be argued that people who built man-made ponds were better managers and thus had higher milk quality.

BEEF CATTLE

Beef cattle includes all animals of the genus *Bos* and their close relatives that are raised primarily for meat production. As ruminants, beef cattle are capable of utilizing a wide range of feedstuffs and consequently are maintained in an array of situations ranging from extensive grazing to confined feedlot pens and intensive laboratory environments. Regardless of the housing system, basic needs for food, water, shelter, and comfort should be met.

FACILITIES AND ENVIRONMENT

Thermal Indices

At the present time, the temperature-humidity index {THI; THI = 0.8 × ambient temperature + [(per cent relative humidity/100) × (ambient temperature–14.4)] + 46.4} has become the de facto standard for classifying thermal environments in many animal studies and selection of management practices during seasons other than winter.

The THI, first proposed by Thom has been extensively applied for moderate to hot conditions, even with recognized limitations related to airspeed and radiation heat loads. A THI between 70 and 74 is an indication to producers that the potential for heat stress in livestock exists. In particular, when THI values are above 70 by 0800h, it is recommended that managers of confined cattle that have high metabolic heat loads initiate or prepare to initiate heat-stress management strategies before cattle become exposed to the excessive heat load.

A THI of 84 or above can cause death, especially in feedlot cattle that are within 45 d of slaughter and consuming high-energy finishing diets. Modifications to the THI have been developed to overcome the shortcomings related to the lack of airflow and radiation heat load in the index. Eigenberg also developed similar adjustments based on predictions of respiration rates using ambient and dew point temperature, windspeed, and solar radiation. These models have merit in that the combined effects of multiple environmental factors can be taken into account when determining animal comfort. Gaughan developed a more extensive index as a guide to the management of feedlot cattle during hot weather.

The heat load index (HLI) incorporates black globe temperature, relative humidity, and windspeed. A threshold above which cattle are less efficient at dissipating heat was developed for a reference animal. The threshold for a full-blood Brahman steer is 96. Also, adjustments to the threshold are possible for use of shade, clean dry pens, cattle coat colour, and days on feed. The thresholds are lowered if cattle are sick or not acclimated to summer conditions. Very limited data exist for assessing environmental effects on reproduction.

However, Amundson found THI and daily minimum temperature to be equally good predictors of pregnancy rate at 42 d into the breeding season. However, the combination of wind speed and THI had the greatest correlation to pregnancy rate. Indices for cold stress are not as well defined as for heat stress. The wind chill index (WCI) has traditionally been used to derive an apparent temperature for humans. In 2001, the National Weather Service released a new WCI that may have merit for assessing effects of wind on domestic livestock.

Ideal Thermal Conditions

Under most environmental conditions, temperature represents a major portion of the driving force for heat exchange between the environment and an animal. However, moisture and heat content of the air, thermal radiation, and airflow also affect total heat exchange. Thus, a combination of environmental variables contributes to the conditions to which an animal responds. Under conditions in which relative measures and comparisons of the effect of different environmental variables could be determined, the apparent ambient temperature at which animals can cope has been defined with a reasonable degree of accuracy; however, variation does exist among animals. Environmental conditions that provide maximum comfort and require little or no energy expenditure for maintenance depend on cattle age, metabolic size, and/or body mass and surface area. The TCZ generally ranges between 15 to 25°C for most cattle less than 1 mo old; between 5 and 20°C for a mature beef cow consuming a maintenance diet; and between–10 and 20°C for yearlings with ad libitum access to energy-dense feedlot diets.

Based on physiological responses and heat load thresholds, *Bos indicus* and some heat-tolerant *Bos taurus* cattle breeds have a TCZ at least 5°C greater than typical *Bos taurus* cattle. Encompassing the TCZ is the thermoneutral zone (TNZ). Within the TNZ, an animal can maintain homeostasis through normal physiological and metabolic processes, which may require minimal expenditure of energy when the animal is exposed to conditions outside the TCZ. The TNZ generally ranges between 10 and 30°C for most cattle less than 1 mo old; between–15 and 28°C for a mature beef cow consuming a maintenance diet; and between–35 and 25°C for yearlings with ad libitum access to energy-dense feedlot diets. Even though the upper end of the TNZ for most *Bos taurus* cattle is between 25 and 30°C, for high-producing cattle with high intakes of metabolizable energy the upper limit may be closer to 20°C on sunny days when little or no wind is present.

When given sufficient time, cattle acclimate and adapt to colder or hotter conditions. It should be noted that cattle that are adapted to–35°C may be uncomfortable at 10°C. Thus, the TCZ and TNZ serve only as guidelines to describe the limits within which cattle are comfortable and can adapt to, respectively. Independent of these guidelines, performance standards that indicate a problem with the thermal environment include, in cold weather, shivering, huddling, and loss of body condition/weight; and in hot weather, panting, sweating, and a reduction in feed intake. Primary factors that affect thermal comfort include feed/energy intake and body condition/fat cover.

Range and Pasture Systems

Acceptable systems for grazing beef cattle on pasture and rangeland vary widely. Cow body condition is an excellent performance standard for monitoring the well-being and nutritional status of range cattle. Special consideration needs to be given to environmental factors that affect grazing beef cattle. In areas where heat stress is common, provision of shade to decrease the solar heat load is the most practical intervention in pasture and range systems.

The need for artificial shade should be assessed after careful consideration of the adequacy of naturally occurring sources. Heat stress is evidenced when respiration rates begin to increase. Prolonged increases in body temperatures will result in decreased feed intake, body condition, and weight. In areas where exposure to extreme cold is likely, provision of shelter for grazing beef cattle may be desirable. Grazing beef cows decrease grazing time and forage intake as ambient temperature decreases below 0°C, although such changes are small in adapted beef cows. Cattle use windbreaks to decrease wind chill and prevent exposure to blowing snow, although it has not been clearly established that windbreaks improve animal performance.

Supplementary feed should be provided during periods of heavy snow cover that preclude grazing. An adequate supply of forage should be available to grazing cattle. Intake and performance may be decreased when the amount of standing forage is lacking but the appropriate quantity of forage dry matter per hectare varies with the pasture or range type and the stocking rate. Guidelines for acceptable amounts of standing forage per unit of body weight at given stocking rates are available but additional research is needed with a variety of pasture and range types. Grazing beef cattle should be provided with supplements for nutrients that are known to be deficient in pasture and range forage in particular localities. In almost all grazing environments, range cattle require free-choice access to supplemental salt as a source of supplemental sodium.

Typically, these salt-based, free-choice mineral supplements will also be fortified with trace minerals. Observation and monitoring of range cattle often occur less regularly than for other livestock. When supplemental feed is provided, cattle are usually observed at least 2 or 3 times weekly. Unsupplemented cattle on open range may be observed less frequently. However, it is recommended that range cattle be observed at least once per week. In certain areas, grazing beef cattle may be affected by predators and poisonous plants. Careful attention should be given to such problems, and efforts should be made to decrease or eliminate these adverse conditions. Availability of fresh, unfrozen water is critical for grazing beef cattle, and distance to water should be given consideration in pasture and range systems.

If cattle are required to travel long distances to water in hot, dry climates, animal performance and utilization of pasture forage can be affected. Holechek recommended that distance to water be no greater than 1.6 km in rolling, hilly country and in undulating, sandy terrain. This recommendation was decreased to 0.8 km in rough country, increased to 2.4 km in smooth, sandy terrain, and increased to 3.2 km in areas with flat terrain. Thus, the distance to water for grazing cattle should not exceed 3.2 km, and every animal should have the opportunity to drink ad libitum at least once per day.

Feedlot and Housing Systems

Beef cattle used in research or teaching may be housed in intensive management systems, either indoors or in open lots, with or without shelter. Facilities for beef cattle should provide cattle with opportunities for behavioural thermoregulation. Management of dairy beef is similar to other cattle, although, some feeding, housing, and marketing regimens are unique to Holsteins. Proper airflow and ventilation are essential in intensive facilities. In feedlots, cable or wire fencing has minimal effect on natural airflow in summer. However, high airflow rates are undesirable during periods of low temperature, and tree shelterbelts and other types of windbreak can decrease the rate of airflow past the cattle. An 80 per cent solid windbreak 3 m high decreases wind speed by half for about 45 m downwind and controls snow for about 8 m; a similar windbreak 4 m high decreases wind speed by half for about 65 m downwind and controls snow for about 10 m.

A windbreak is recommended in mounded, south-sloping feedlots in the northern United States to provide dry resting areas with low air velocities.

Caution should be exercised when placing cattle in sheltered areas in the summer because of the adverse effects of restricted airflow on cattle reared in hot environments. During potentially stressful heat episodes, panting scores (1 = elevated respiration rate, 2 = drool or saliva present on side of mouth, 3 = open mouth breathing observed, and 4 = tongue and neck extended with open mouth breathing) can be utilized as an excellent indicator of stress levels experienced.

When cattle are beginning to experience panting scores of 2 or greater some means of cooling may be needed. Cattle learn to take evasive action to alleviate heat stress and such competition for cooler areas in a pen or around the water trough increases even during cooler days in which heat alleviation methods are not utilized. When this occurs evidence of crowding is observed, which exacerbates the heat stress problems.

Wetting the ground or floor of holding facilities can be an effective method of cooling cattle managed in unshaded, outdoor units where surface vegetation is sparse or non-existent. Direct wetting of cattle during extreme heat is also an effective practice and is often used as an emergency measure. Benefits of sprinkling are enhanced if sprinkling is started in the morning, before cattle experience high heat loads. Generally, a daily application of 0.5 to 1.0 cm of water is sufficient to cool pen surfaces. However, applying 1.25 to 1.50 cm every other day is acceptable and will not sufficiently contribute to mud build-up in normally dry pens.

In areas with high evaporation rates additional water may be needed, which can serve to cool pen surfaces as well as eliminate potential dust problems. The size of the area needed to be sprinkled would be similar to the shade area recommendations. As a routine protective practice, wetting can be efficiently accomplished by utilizing a timer to provide 5 to 10 min of spray during each 20-to 30-min period. Fogger nozzles are often mistakenly recommended for wetting animals.

Fogger nozzles are less effective than sprinkler nozzles because of the barrier formed by the fine droplets. These droplets adhere to the outer hair coat of the animal, causing the heat for evaporation to come from the air rather than from the body.

Mitlöhner *et al.* reported that misting cattle was not as effective as shade in decreasing heat stress, and in some cases, caused respiration rate to increase compared with non-misted cattle. Shade for cattle can provide the margin of survival for animals that are unconditioned to a sudden heat wave with high solar radiant loads in central and southern regions of the United States. Mader found limited performance benefits of utilizing shade in the north-central region of the United States, in contrast to the findings of Mitlöhner where shade was effective in southern regions.

Also, use of shade in northern climates may be costly and logistically prohibitive because of snow load requirements, potential mud problems under shade and the low percentage of time that cattle may actually benefit from using the shade. However, benefits of using shade for maintaining animal comfort will almost always be found in any area or location in which abnormally hot or hot and humid conditions arise or persist, including northern climates, and when cattle have not had the opportunity to acclimate. Mitlöhner *et al.* found excellent results when shades were provided for feedlot cattle reared in the south-central region of the United States, an area where more consistent benefits of shade would be expected to be realised. For optimum benefits shades should be 3.6 to 4.2 m high in areas with clear, sunny afternoons to permit maximum exposure to the relatively cool northern sky, which acts as a radiation sink.

In areas with cloudy afternoons, shades 2.1 to 2.7 m in height are more effective, as they limit the diffuse sky radiation received by animals beneath the shades. The amount of shade required for young cattle is 0.7 to 1.2 m^2 per animal, whereas larger cattle need 1.8 to 2.5 m^2 per animal.

Shades are strongly recommended for sick cattle or for animals in hospital pens. Cold housing can be provided for beef cattle. Open sides of any cattle building need to face away from prevailing winds. Such structures are ventilated by natural airflow, and the resultant winter temperatures are typically 2 to 5°C above outdoor conditions as a result of body heat. Totally enclosed housing requires ventilation to maintain the air temperature at acceptable levels and to minimize the accumulation in the air of water vapour, noxious gases, other odourous compounds, and dust. Ventilation systems may be either natural or mechanical.

Type of pen surface affects dustiness during hot dry weather and mud or manure build-up during wet periods. Good drainage of outside pens is imperative. Dirt pens should be regularly cleaned of animal waste residues and maintained to minimize accumulation of water. A hard surface apron in front of the feed bunks and around water troughs and shelters should be considered in dirt pens. Mounds should be provided in dirt pens for cattle to lie on during inclement weather.

Accumulation of mud in a pen or on the cattle can influence maintenance requirements and thermal balance. Properly designed pens with adequate slope are extremely important for minimizing mud and related health and behaviour problems. In areas where slope or drying conditions are limited, adding mounds is very useful for keeping cattle clean and dry. Under hot-humid conditions, mounds aid in preventing animal crowding and improve exposure to airflow for the animals that utilize them. Additional information on feedlot/drylot pen design and layout has been published by Pohl and Henry. For hard-surfaced pens, materials should be durable, slip-resistant, and impervious to water and urine; easily cleaned; and resistant to chemicals and corrosion from animal feed and waste. Concrete floors should be scored or grooved during construction to improve animal footing.

Properly designed slotted floors are self-cleaning. Fences, pen dividers, walls, gates, and other surfaces must be strong enough to withstand the impact of direct animal contact. Configuration and treatment of contact surfaces must minimize or eliminate protrusions, changes in elevation, and sharp corners to minimize bruising and injuries and to improve the efficiency of cattle handling. Proper lighting permits inspection of animals in feedlots and other cattle housing systems and provides safer working conditions for animal care personnel. Maintenance of facilities should be timely and ongoing.

FEED AND WATER

Diets for beef cattle should be formulated according to the recommendations of the NRC. Formulation of diets should consider factors such as environmental conditions, breed or biological type, sex, and production demands for growth, gestation, or lactation. Feed and water should be offered to cattle in ways that minimize contamination by urine, feces, and other materials. Feed bunks should be monitored daily and contaminants or spoiled feed should be removed. In most situations, feed should be available at all times.

However, restricted feeding of high-energy diets may be practiced to meet maintenance requirements or targeted levels of production. When restricted feeding is practiced, feed must be uniformly distributed in the bunk to allow all cattle to have simultaneous access to the diet. When high-energy diets are fed, increased attentiveness should be given to possible occurrence of diet-related health problems such as grain overload, lactic acidosis, and bloat. Abrupt changes in diets should be avoided. Feed deprivation for more than 24 h should be avoided, and feed deprivation for any length of time must be justified in the animal use protocol. Cattle can vary considerably in body weight and condition during the course of grazing and reproductive cycles. Feeding programmes should allow animals to regain the body weight that is lost during the normal periods of negative energy balance.

Confined cattle should have continuous free access to a source of water, except before surgery or weighing if the research or animal care protocol requires such restriction. When continuous access to water is not possible, water should be available ad libitum at least once daily and more often if hot weather conditions exist or cattle have high levels of metabolizable energy intake for purposes of achieving high output. Under winter range conditions, Degen and Young found that snow can be used as a water source for beef cows and growing calves. However, there was evidence that the snow resulted in reduced water intakes as evidenced by compensatory water intake when water was reintroduced following 84 d of consuming water in the form of snow. When snow was the only source of water, total water intake reductions averaged approximately 10 per cent among the cattle groups.

The quantity and, possibly, quality of water available will influence animal comfort, especially under hot conditions. Evaporation of moisture from the skin surface or respiratory tract is the primary mechanism used by the animals to lose excess body heat in a hot environment. Estimates of daily water requirements for beef cattle are reported in NRC.

During summer months, in particular, waterer space available and water intake per animal becomes extremely important. Under these conditions, Mader found that as much as 3 times the normal waterer space may be needed to allow for sufficient room for all animals to access and benefit from available water. Additional waterer space recommendations are provided by MWPS.

ADEQUATE CARE OF CATTLE AND CALVES IN HUSBANDRY

Adequate care of cattle and calves is especially important for establishing and maintaining optimal immune system function. Good husbandry can minimize health problems and infectious diseases. The risk of disease and mortality in young calves is related to immune status.

It is critical that newborn calves nurse or ingest colostrum soon after birth. The health of young growing cattle should be assessed regularly pre-and postweaning. Animal care personnel should be taught to recognize signs of illness and external parasites.

Alert caretakers should have the ability to perceive appropriate behaviour and posture. A system of monitoring calves through critical stress periods such as weaning should be established. Any sick or injured calves should be treated promptly. Daily records should be kept. For cattle reared in close confinement assessments should be done at least once daily and more often if cattle have been stressed or potentially exposed to conditions in which their health could be compromised.

In general, confined feedlot cattle, especially new incoming cattle, require more frequent observations than non-confined cattle because of the greater probability of animal health being compromised due to comingling, dehydration, digestive problems, respiratory problems, and interaction of any of these factors with environmental stress. Signs of healthy calves are alert ears and clear eyes, no signs of diarrhea, and, upon arising, resumption of a normal standing posture after stretching. For feedlot cattle provided energy-dense diets, caretaker knowledge of acidosis and management regimens necessary to minimize digestive problems are essential. Appropriate medication and vaccination programmes should be used to reduce the incidence of disease and mortality, improve cattle health and performance, and ensure that no illegal residues occur in the carcass. Treatment and vaccination schemes should be based on veterinary advice and experience.

Weaning

In typical beef cow/calf production systems, calves are artificially weaned from their dams by physical separation. This process, albeit important to the efficiency of the cowherd, can be stressful to both the cow and calf. The most common weaning procedure involves an abrupt separation of cows and calves resulting in increased walking and vocalization and decreased eating and resting. An alternative to abrupt weaning and permanent separation is a period of fenceline contact between cows and calves in adjacent but separate pastures.

This weaning management alternative has been shown to decrease vocalization and walking and increase the time spent resting and grazing. This fenceline weaning procedure may also decrease the incidence of calf illness. Within the weaning pasture or pen, a mature cow can be included in the group of freshly weaned calves. This "trainer" cow can assist in introducing the weaned calves to the location and facilitating consumption of feed and water.

Despite the weaning process selected, it is important that weaned calves be provided access to clean water and a source of feed and/or forage. To encourage intake, highly palatable forage and feed sources are recommended until calves become accustomed to the separation from their dams. Additionally, feed and water sources should be placed close to the perimeter of the fenceline, because calves will typically spend a majority of their time in these areas as they seek to reunite with their dams.

Social Environment

Cattle are social animals. Each individual in the group should have sufficient access to the resources necessary for comfort, adequate well-being, and optimal performance. Mixing, crowding, group composition, and competition for limited resources are part of the social environment and in some circumstances, may be social stressors for certain cattle.

Generally, cows from similar environments but from different social groups can be mixed with little or no long-term adverse effect on performance; however, because introduced cows may be the recipients of aggression, the number of mixing episodes should be minimized. Mixing of older cattle, especially bulls, results in more fighting than occurs when younger cattle are mixed. Fighting and mounting can be a problem associated with keeping bulls in social groups and can present a significant welfare problem if not managed carefully. Attempts should be made to keep bulls in stable social groups and to minimize mixing. When feed, water, or other resources critical for comfort or survival are limited, or when large differences exist among cattle in size or other traits related to position in the social order, some animals may be able to prevent others from gaining access to resources.

In properly designed facilities, all individuals should have sufficient access to feed, water, and resting sites to minimize the correlation between position in the social order and productive performance. Proper animal care includes observation of groups and of individuals within groups to ensure that each individual has adequate access to the resources necessary for optimal comfort, welfare, and performance.

Floor or Ground Area

Area recommendations for open lots and barns are listed in Table. Every animal should have sufficient space to move about at will, adequate access to feed and water, a comfortable resting site, and the opportunity to remain reasonably dry and clean. These suggested recommendations alone do not ensure that an ideal environment exists; however, in some cases these conditions can be met with less than the recommended area.

The area required is affected by type and slope of floor or soil surface, amount of rainfall, amount of sunshine, season, group size, and method of feeding. Open feedlot pens need to be sloped to promote drainage away from feed bunks, waterers, pen dividers, and resting areas. Space allocations are related directly to slope. In temperate Midwestern climates, the following relationships have been found to be workable: 2 per cent slope or less: 37 to 74 m^2 per animal; 2 to 4 per cent slope: 23 to 37 m^2; and 4 per cent or greater slope: 14 to 23 m^2. Space allocations can be less in drier regions of the country. In the Southwest, at 0 per cent slope, typical allocations are 14 to 23 m^2 per animal. In other regions, space allocations may need to be increased above Midwestern norms in consideration of such factors as soil type and rainfall distribution. The area requirements for cattle are greatly influenced by group size. One animal housed separately in a pen requires the greatest amount of floor area on a per-animal basis. As group size increases, the amount of area required per individual decreases.

When an animal is housed individually, the minimum pen width and length should be at least equal to the length of the animal from nose tip to tail head when the animal is standing in a normal erect posture. Acceptable indoor pen floor surfaces for beef cattle include unfinished concrete, grooved concrete, concrete slats, expanded metal, plastic-covered metal flooring, and rubberized mat. The floor surface in stanchions and metabolism stalls may be concrete, expanded metal, wood, rubberized mat, or a combination of materials that provides support for the animals' bodies; does not damage hooves, feet, legs, and tails; and can be cleaned.

STANDARD AGRICULTURAL PRACTICES

For beef cattle, management procedures may be performed by properly trained, non-professional personnel. These include, but are not limited to, vaccinating, dehorning and castrating young cattle, horn-tipping, ear-tagging, branding, weighing, implanting, use of hydraulic and manual chutes for restraint, roping, hooftrimming, routine calving assistance, ultrasound pregnancy checking, feeding, and watering.

Other husbandry and health practices used in beef cattle research and teaching that similarly may be performed by properly trained, non-professional personnel, but that require special technical training and advanced skill levels, include artificial insemination, electroejaculation, pregnancy palpation, embryo flushing and transfer, non-routine calving assistance and dystocia treatment, emergency cesarean section, retained placenta treatment, and dehorning and castration of older cattle.

One of the main animal husbandry concerns is that of pain and distress, especially pain inflicted from standard husbandry procedures. Dehorning, castration, and branding are husbandry procedures that can cause pain and discomfort; nevertheless, these procedures are justified as a management tool to minimize injuries or other problems associated with confining horned cattle and commingling bulls.

Dystocia Management

Matings should be planned to lessen the genetic probability of dystocia. When dystocia does occur, proper care and assistance at calving can decrease injury or death of both calves and heifers/cows. Parturition without complication is common in beef cows. Therefore, before administering assistance to a cow experiencing difficulty with calving, personnel should be familiar with the stages associated with approaching parturition and the signs of normal delivery.

As a general rule, females should be examined within 30 to 60 min following presentation of feet, nose, or fetal membranes if delivery of the calf does not appear imminent. However, heifers or cows exhibiting signs of a malpresentation, oversized fetus, fetal anomaly, or other obvious complication must be assisted immediately. Facilities should be provided that are designed for restraint of cows and heifers experiencing dystocia.

Because many animals, especially heifers, lie down during the obstetrical procedure, sufficient space should be provided to permit adequate freedom of movement. It is important that the obstetrical restraint facility be fitted with side gates, both of which are hinged at the head end, so that the animal can become fully recumbent and the obstetrical procedure can be performed with safety and efficiency. In dystocia cases where fetal presentation appears to be compromised or there appears to be a disparity between the size of the fetus and the diameter of the birth canal, assistance of the delivery by personnel appropriately trained in the judicious use of a fetal extractor may be attempted. In general, if more than slight traction is required on the fetal extractor, the procedure should be stopped and a veterinarian called immediately to perform a caesarean section or fetotomy. Use of excessive force can damage the calf and/or dam and lead to suffering and/or death. Strict sanitation should be used with all obstetrical procedures.

Vaccinations and Drug Administration

Vaccinations are a key component to any herd health programme. Care should be taken to ensure the proper use, handling, and storage of vaccines and approved or investigational drugs. The preferred site of injection is the neck for either intramuscular or subcutaneous injections; however, for investigational drugs used in research, alternate sites of administration may be required or preferred as dictated by the research protocol. Investigators and animal care staff should utilize best management practices associated with the use of syringes and handling needles. Use and regular replacement of disposable syringes and needles is highly recommended to avoid excessive trauma and disease transmission.

Castration

Castration of male beef cattle is performed to reduce aggressiveness, prevent physical danger to other animals in the herd and to handlers, enhance reproductive control, manage genetic selection, and satisfy consumer preferences regarding taste and tenderness of meat. Castration of young bulls is a necessary management practice in beef production.

Several methods for castrating cattle are acceptable, including surgical removal of the testicles using a knife or scalpel to open the scrotum and cutting or crushing the spermatic cords with an emasculatome or emasculator. Bloodless procedures utilizing specialized rubber rings or surgical tubing bands are available to create devitalization and eventual sloughing of the tissues below the ring or band. High-tension banding systems may be used with appropriate veterinary supervision and/or training in those situations where surgical castration may predispose to postsurgical complications or when surgical castration is not appropriate because of its effect on research protocol. The castration method used should take into account the animal's age and weight, the skill level of the technician, environmental conditions, and facilities available as well as human and animal safety.

Whatever the method of castration, the procedures should be conducted by, or under the supervision of, a qualified, experienced person and carried out according to castration equipment manufacturer recommendations and accepted husbandry practices.

Surgical castration is normally a short-term event with short-term duration of pain-associated responses. Bloodless castration has been associated with lower short-term pain indicators but longer chronic pain indicators. Bloodless castration should be used when surgical castration may predispose to postsurgical complications or when surgical castration is not appropriate because of its effect on the research protocol.

Castration is least stressful when performed at or shortly after birth, but lower stress is reported if performed before 2 or 3 months of age or before animals reach a body weight of 230 kg. It is strongly recommended that calves be castrated at the earliest age possible. It may be desirable to inject local anesthetic in the scrotum of calves heavier than 230 kg when surgical methods of castration are used or when the spermatic cords are crushed.

Topical local anesthetics may also be used on open wounds. Improved animal performance, as one potential indicator of improved animal welfare, has not been observed in animals locally anesthetized at the time of castration. It should be recognized that the effect of anesthetic agents is short-lived. Nevertheless, procedures should be implemented to minimize pain and discomfort, especially in older cattle.

Castration of older, heavier bulls should be performed only by skilled individuals. When it is necessary to castrate these heavier bulls, techniques and procedures to control bleeding must also be used. No advantage to use of anesthesia is apparent when bloodless castration is practiced. The possibility of infection should be given additional consideration after castration. Equipment should be sterilized, and facilities should be clean and sanitized. Infection following castration can be minimized by keeping the animals in a clean area and away from excessive mud or contaminants following the procedure until the wound is healed.

If tetanus is a common disease associated with the premises, or if a bloodless castration method is utilized, the herd health veterinarian should schedule a prophylactic tetanus immunization programme.

Dehorning

Horns on cattle can cause bruises and other injury to other animals, especially during transport and handling. Horns on adult cattle also can be a hazard to humans. Hornless cattle require less space in the feedlot and at the feed bunk. Polled breeds should be used whenever possible.

Disbudding and dehorning of cattle in the United States is not currently regulated. The Canadian Veterinary Medical Association recommends that disbudding be performed within the first week of life. In the United Kingdom, disbudding with a hot iron is preferred to dehorning and it is advised that this should be performed before cattle reach the age of 2 mo. In Australia, dehorning without local anesthesia or analgesia is restricted to animals less than 6 mo old. Calves suffer less pain and stress, have less risk of infection, and have better growth rates when dehorning is performed at a very young age. Stafford and Mellor found that the use of local anesthetics virtually eliminated the escape behaviour of calves associated with the dehorning process and that a 2-h delay was observed in the cortisol response to horn amputation. Whenever possible, the use of a local anesthetic is encouraged when dehorning.

When horned breeds of cattle are selected, dehorning should be performed under the supervision of experienced persons using proper techniques. The horn buds should be removed at birth or within the first month after birth by several means, including hot cauterizing irons, cauterizing chemicals, a sharp knife, or commercially available mechanical devices. It is strongly recommended that calves be dehorned at the earliest age possible. When it is necessary to remove horns from older cattle, methods that minimize pain and bleeding and prevent infection should be employed.

Dehorning should be performed by a person knowledgeable and experienced in the appropriate procedures. Appropriate restraint and local anesthesia to control pain should be used when cattle older than 1 mo of age are dehorned. Cattle should be monitored for hemorrhage and infection following dehorning. Adult cattle should be dehorned if aggressive behaviour is displayed towards herd mates or humans. Dehorning may temporarily depress the growth of cattle. In the event that bunk and pen space are ample, then tipping the horn may be considered as an alternative to minimize potential bruising or injury of pen mates. However, Ramsay reported that, after transport, carcass bruises were as common among tipped cattle as among horned ones.

Identification Methods

Proper animal identification is essential to research, facilitates record keeping, and aids in the routine observation and repeat identification of cattle. Methods of identification include skin colour markings, ear tagging, tattooing, hot branding, freeze branding, and electronic identification. Ear tags are best used in conjunction with a more permanent form of identification such as a tattoo or brand, as ear tags are sometimes lost. Hot branding the hide is utilized as a means of identification; however, loss in hide value and studies indicating that freeze branding is less painful than hot branding have begun to minimize the use of hot branding. Alternatives to hot branding should be considered. However, skin and hair colour in addition to a limited access to liquid nitrogen or dry ice in extensive range operations may affect the ability to achieve a quality freeze brand.

At some locations, branding is required by law. Both hot branding and freeze branding should be performed by trained personnel to minimize skin contact with the branding device to only that required to achieve a useful brand. Advent of a national animal identification system (NAIS) in the form of visual or *radio frequency identification* (RFID) ear tags serve as an additional means of identification. As this system will become standard for all cattle as part of a national programme, managers of beef cattle as part of resident herds used in research should comply with the established guidelines.

Implanting

Implanting of cattle is a management practice for the administration of growth promotants and potentially as a means of delivery of investigational compounds used in research. For proper absorption and maximum response, implants should be placed correctly and in the correct location. Traditionally, implants are placed beneath the skin on the back side of the middle third of the ear; however, alternate implantation sites may be required as designated by the research protocol.

Proper disinfection of the implant site is required to prevent infection. Care should be taken not to injure major blood vessels or the cartilage of the ear when implanting in the ear location. Utilization of best management practices associated with the use of the implant device and correct needle-handling procedures are required by suitably trained personnel.

SPECIAL CONSIDERATIONS AND INTENSIVE LABORATORY FACILITIES

Some research and teaching situations require that beef cattle be housed under intensive laboratory conditions. Cattle may be kept in metabolism stalls, stanchions, respiration chambers, or environmental chambers. Housing cattle in such facilities should be avoided unless required by the experimental protocol and then should be for the minimum amount of time necessary to accomplish the teaching or research objective.

Cattle that are held or penned temporarily in crowded areas, frequently disturbed, or come into close contact with humans, or exposed to unfamiliar conditions or laboratory/teaching settings should have calm dispositions and be adapted to frequent contact with animal care personnel and to those conditions that could result in the animal having an adverse reaction. In some cases, it may be advantageous to train such animals to a halter. Time spent preparing cattle for use in a laboratory improves the quality of research and the safety of both the animals and the humans. Cattle should not be housed in isolation unless approved by the Animal Care and Use Committee for specific experimental requirements. Whenever possible, cattle should be able to maintain visual contact with others. Unless the experimental protocol has special requirements for lighting, all animal rooms should be designed to minimize variation in light intensity.

During light periods, the minimum light intensity for intensively housed cattle is 70 lx. If possible, a diurnal light-dark cycle should be used and a standard daily schedule established. Excreta should be removed from enclosed laboratories at least once daily. Pens or stalls should be washed thoroughly at the beginning of every trial. If excreta or other foreign materials such as wasted feed cannot be adequately removed through daily cleaning, additional washing may be needed during a trial.

The method of collection of feces and urine from cattle in metabolism stalls, stanchions, and chambers depends on the design and construction of the unit. Additional management may be needed to keep animals clean when they are housed in stalls or stanchions.

Cattle may need to be washed and curried regularly to maintain cleanliness and to avoid fly infestations. Pens, stalls, and stanchions should be large enough to allow cattle to stand up or lie down without difficulty and should be long enough to allow cattle to maintain a normal standing position. Because of the operating costs associated with singlepass ventilation systems in controlled environmental facilities, partial recirculation of exhaust air from animal rooms is common and acceptable in many studies.

In facilities designed to recirculate even a small part of the exhausted air, treatment is necessary to remove odourous compounds, gases, and particulate matter. Cattle maintained in some laboratory environments have their activity restricted more than cattle in production settings. The length of time that cattle may remain in stanchions, metabolism stalls, or environmental chambers before removal to a pen or outside lot for additional exercise should be no longer than that necessary for conducting the study.

Opportunities for regular exercise should be considered if they do not disrupt the experimental protocol; care must be taken in moving animals from the laboratory to the outside environment for exercise when a large temperature differential exists. If cattle are to be housed in such laboratory environments for more than 3 wk then particular attention should be given to alertness of the animal; appetite; fecal and urinary outputs; and condition of the feet, legs, and hock joints. Rubber mats or suitable alternatives should be used to increase the comfort of cattle maintained for lengthy periods on hard surfaces.

Use of Beef Cattle in Biomedical Research

Relative size, cost of maintaining beef cattle, and the use of alternate animal models in biomedical research have largely minimized the use of beef cattle in this regard. Nevertheless, beef cattle have played a role in understanding such maladies as lysosomal storage diseases and hemochromatosis, among others, which have similarities to diseases, often genetically based, found in humans; therefore beef cattle may serve as highly valuable biomedical models in some cases.

In addition, the potential use of cattle as bioreactors for the production of human gene products or pharmaceuticals in milk, blood, urine, or tissues may further extend the use of beef cattle for biomedical applications. Standards for the care and welfare of beef cattle used in biomedical research should be the same as that applied to all beef cattle. However, institutional or biomedical funding agencies may require more specific disease entry testing requirements for cattle used in biomedical research, in addition to having more stringent procedures with respect to adherence to alternate oversight committee guidelines for reporting, housing, observation and care procedures than might be utilized or generally accepted under typical agricultural research and production systems.

In some cases, in which in vitro reproductive technologies are used for the production of beef cattle in research, maturation, fertilization, manipulation, and/or culture, differences can exist in fetal morphology, physiology, and in the expression of developmentally important genes that may require alteration in management strategy. For example, cattle produced in this manner may exhibit "large calf syndrome" and therefore may require extra assistance at calving. The animal biotechnology sector continues to grow, with significant advancements being made that may directly and indirectly affect beef cattle research, and it is important to recognize that alterations through the genetic engineering of beef cattle may similarly require alterations in beef cattle care practices. With respect to genetic engineering, unanticipated results from genetic modifications have been observed in several genetically engineered species that require diligence on the part of the researcher and animal care staff in assessing animal welfare. However, the general standards of care associated with genetically engineered and cloned beef cattle should be the same as that applied to all beef cattle unless the specific genetic modification requires an alteration in management within the research environment to specifically facilitate animal welfare.

EUTHANASIA

The USDA and Food Safety and Inspection Service (FSIS) *Humane Slaughter of Livestock* regulations, floors of livestock pens, ramps, and driveways of harvest facilities shall be constructed and maintained so as to provide good footing for livestock. Animals shall have access to water in all holding pens and, if held longer than 24 h, access to feed. Also, for animals held overnight there shall be sufficient room in the holding pens for the animals to lie down.

The AVMA *Guidelines on Euthanasia* lists several methods of euthanasia that are appropriate for ruminants. Intravenous administration of barbiturates, potassium chloride used in conjunction with general anesthesia, and penetrating captive bolt are acceptable means of euthanasia in all cases. Other conditionally acceptable methods include intravenous administration of chloral hydrate, gunshot to the head, and electrocution.

In all cases, euthanasia should only be performed by trained individuals. Agents that result in tissue residues cannot be used for the euthanasia of ruminants intended for human or animal food, unless those agents are approved by the Food and Drug Administration. Carbon dioxide is the only chemical currently used in euthanasia of food animals that does not lead to tissue residues. Use of carbon dioxide is generally not recommended for euthanasia of larger animals. The carcasses of animals euthanized by barbiturates may contain potentially harmful residues, and such carcasses should be disposed of in a manner that prevents them from being consumed by humans or animals.

Dying, diseased, and disabled livestock shall be provided with a covered pen sufficient to protect them from adverse climatic conditions. Incurably ill or injured animals in chronic pain or distress should be humanely euthanized as soon as they are diagnosed as such and according to AVMA recommended procedures. Their disposal should be accomplished promptly by a commercial rendering service or other means according to applicable ordinances and regulations.

8

Managing Manure on Organic Farms

LIVESTOCK MANAGEMENT PRACTICES

To maximize forage productivity, it is best to combine or alternate various livestock breeds on a range. Their differing food preferences can help to keep plants productive by minimizing overgrazing of a particular favoured area and allowing less preferred plant species time to mature. It is prudent to make superior forage available to those animals with the highest needs. When forage is limited, livestock managers may decide that young and milk-producing animals must have first access to new pastures and ranges with a wide variety of abundant forage.

Managers should investigate the value of various systems of rotating livestock. Rotation allows land to be grazed continuously throughout the year. Livestock can be rotated between fields or ranges to prevent the buildup of disease and to vary grazing pressures. Through either fencing or herding, they can be relocated into croplands to consume crop residues. Assessment of seasonal grazing patterns should include potential impact on soils.

Dry-season grazing can benefit the land by breaking up crusted soil and working seeds into the ground. By contrast, grazing on moist soil can cause considerable soil compaction, which reduces the soils ability to absorb moisture and can result in increased erosion from run-off during the rainy season. Many of the environmental impacts from livestock production are associated with particular practices of livestock managers.

Thus it is critical to understand current practices and how the proposed project might alter these practices or promote new ones:

- Who are the local community's livestock managers?
- What practices do a family or community use to control the size and composition of livestock herds?
- How do livestock managers currently control livestock movement? Will the proposed project change these movements in a way that might harm the environment?

- Does the proposed project require the construction of fences? If so, will they interfere with wildlife migration or transit of livestock belonging to other communities? Could the fences lead to overgrazing and land degradation? Will the fences be built with local materials? Would living fencing be practical? Would solar-powered electric fencing be technically and economically feasible?
- Are streams and riverbanks currently protected from livestock damage? If the proposed project will open new areas to grazing, will water supplies need to be protected?
- Must steps be taken to prevent new livestock and associated animals (*e.g.*, dogs) from transmitting disease to wildlife? Is there a vaccination/animal disease control programme available for this purpose?
- Will the project involve construction of improvements (*e.g.*, boreholes or other infrastructure)? Could these lead to unplanned changes in herding patterns and overgrazing?

ASSESS DEMAND AND USE OF LIVESTOCK PRODUCTS

- Who is marketing livestock and livestock products?
- Is the demand for livestock products coming from local or outside populations? How rapidly is it increasing or decreasing? How stable is the demand?
- In preparing livestock products, are people using technologies which reduce impacts on the environment, open additional markets, or improve health and nutrition (of people and animals.)

THE LIVESTOCK ECONOMY IN DEVELOPING COUNTRIES

Livestock is becoming increasingly important in the growth of agriculture in developing economies. The contributions made by livestock to both agriculture and gross domestic product (GDP) have risen, at a time when the contribution of agriculture to GDP has fallen. The demand for livestock products is a function of income, and sustained growth in per capita income, rising urban populations and changes in diet and lifestyle are fuelling growth in livestock production.

Livestock production contributes to socioeconomic development in many ways, by augmenting income and employment and reducing the incidence of rural poverty. Though the role of livestock in ensuring nutritional security is recognised in mixed crop-livestock systems, the importance of livestock goes beyond direct food production.

Livestock supply draught power and organic manure to the crop sector, and hides, skins, bones, blood and fibre are used in many industries. Thus, livestock are an important source of income and employment, helping to alleviate poverty and smooth the income distribution among small landholders and the landless, who constitute the bulk of the rural population and the majority of livestock owners.

In addition, livestock can easily be converted into cash and thus act as a cushion against crop failure, particularly in less favoured environments. By enabling crop residues and by-products to be used as fodder, livestock production contributes positively to the environment.

Animal owners in the developing world are predominantly resource-poor small-scale operators with little or no land and few animals, who must operate within the constraints of the local climate and who have limited purchasing power and little access to resources or opportunity to determine resource allocation for animal production. The situation of the poorest livestock owners is fast deteriorating owing to the fragmentation of limited holdings, exhaustion of land resources and increasing human and animal population pressure.

Low livestock productivity in many developing countries is considered to reflect, among other things, the inadequate supply of animal husbandry and veterinary services. Veterinary services have traditionally been provided by the State, but financial constraints have limited the availability and effectiveness of public services.

THE IMPLICATIONS OF TECHNOLOGY

A major benefit of agricultural research and technology is that the purchasing power of the poor increases, because both average incomes and access to staple food products are improved. Studies by economists have provided empirical support for the proposition that growth in the livestock sector affects the whole economy.

Rapid growth of livestock production has stimulated demand for and increased the value of land, labour and non-agricultural goods and services, thereby leading to overall economic growth. The poor spend a relatively high proportion of any additional income on food, so increases in livestock production achieved through the use of biotechnology can have major nutritional implications, particularly if the technology is aimed at the poorest producers.

However, studies have revealed that the commerci-alisation of agriculture has reduced the nutritional security of the poor. Once production of milk, meat or eggs has been enhanced through the use of technology, it is hoped that it will also make a significant difference in other areas such as nutrition, prevention of diseases, health care and other management practices. It is in these areas that biotechnology shows promise and is currently being used. Green Revolution technologies (*i.e.* those technologies designed to improve the efficiency of agricultural processes and increase crop productivity by relying on the extensive use of chemical fertilizers/pesticides and heavy machinery) are intended to be used in package form (*e.g.* new plant varieties supplied with recommendations on fertilizer, pesticide and herbicide rates and water control measures); however, among livestock producers many components of these technologies have been taken up in a piecemeal, often stepwise, manner.

The sequence of adoption is determined by availability and by the potential cost savings. The sequential adoptions of crop management technologies for rice and wheat have been assessed in detail, but few similar studies have focussed on livestock production in developing countries. Evidence from the People's Republic of China, Mexico, South Africa and India suggests that small farmers have had no more difficulty than larger farmers in adopting the new technologies.

The question, therefore, is not whether biotechnology can benefit small-scale resource-poor farmers, but rather how biotechnology can address the agricultural problems faced by farmers in developing countries. Biotechnology is a promising new tool in the development of applied agricultural technologies. The challenge is to focus this potential on the problems experienced by developing countries. The introduction of multi-enterprise systems or, more broadly, agricultural diversification is seen as the way forward for agriculture in the developing world; such systems could lift small-scale and marginal farmers out of poverty.

For example, rotating rice and wheat cropping with dairy farming yields higher profits. Introducing multi-enterprise systems involving livestock enhances the purchasing power of farmers and helps them to obtain nutritional security.

It also generates rural (both farming and non-farming) employment, thus preventing excessive migration to urban areas, which is a common problem in developing economies. Multi-enterprise systems also support the natural environment and contribute to capital formation, thus leading to higher overall growth in the agricultural economy. The technology, infrastructure and institutions now exist to make the application of biotechnology in the context of a multi-enterprise system involving livestock production economically viable.

Global Advantage from Livestock of Developing Countries

The multiplicity of genes, species, populations and agroecosystems in the developing countries of South and South East Asia, the People's Republic of China, Africa and Latin America is viewed as a valuable resource for the genetic improvement of livestock on a global scale. The livestock in these countries, which are an integral part of a fragile ecosystem, are a rich source of animal biodiversity. Buffaloes, sheep, goats, camels and zebu cattle have adapted to their regional environments over thousands of years and have provided an important source of sustenance for the population of the region.

Livestock production in the developing world has a number of advantages over production in more developed countries, for example:

- The unique and valuable production traits of buffaloes, cattle, sheep, goats and camels
- The low-input production system
- The low unit cost of production
- The lean meat produced from sheep, goats and buffaloes
- The considerable biodiversity
- Animal breeds that are resistant to stress and to particular diseases
- The ability of the animals to survive on high-roughage feeds
- The potential for biopharmaceutical developments to lead to significant benefits
- The potential for expanding the microbial food, feed and leather industries
- The integrated production system tailored to the local ecology
- The potential for integrating knowledge and industry.

Several genes and desirable traits have been identified in the livestock of developing countries in Asia and Africa, and some of the livestock species and breeds from these countries have become major contributors to the economy of South America.

Examples of breeds from the developing world that are particularly important on a global level are:

- Buffaloes that produce milk with a high fat content or with the protein quality required to produce mozzarella cheese
- Goats from cold dry regions that produce pashmina and toos (the finest wool in the world)
- Black bengal goats that carry a gene for high prolificacy
- Garole sheep that carry genes for twinning
- Andaman goats that are highly tolerant of salt
- The yak and mithun that are adapted to high altitude
- The camels, sheep and goats that are adapted to a tropical arid environment and can tolerate feed with a high lignin content
- The many species that are resistant to stress or to particular diseases.

Economic Impact of Technologies

The genetic resources possessed by animals in developing countries often affect economic development. Traxler has discussed the economic impacts of biotechnological innovations, but the research and policy options need separate consideration. Animal biotechnology is the result of a multistage process, involving research, development, testing and registration, production and marketing.

The goal is to develop a technology, process or product that has clear commercial potential and can be commercialised after due testing and regulatory approval. Developing countries find it difficult to develop biotechnology because the facilities or resources needed to complete all of the stages in the process are often lacking. However, several technologies from developed countries have been successfully adopted by developing countries. The impact of technology can be analysed by estimating the growth of total factor productivity (TFP) in livestock production. Not many TFP studies on livestock have been reported.

However, separate TFP estimates for the aggregate crop and livestock sectors have been made. TFP analysis has shown a shift towards larger, more commercial and more intensive production systems and has further revealed that, as specialisation has developed over the past decade, the importance of backyard livestock production has declined and the importance of specialised household and commercial enterprise has increased.

Studies from India have shown that technological input is responsible for about 45 per cent of total output growth and that the TFP growth may be as much as 1.8 per cent.

TECHNOLOGIES THAT HAVE A SPECIFIC IMPACT IN DEVELOPING COUNTRIES

There are a large number of technologies that have been developed for or adapted to the livestock of both developed and developing countries. However, the major technologies that are used effectively in livestock production in the developing world include conserving animal genetic resources, augmenting reproduction, embryo transfer (ET) and related technologies, diagnosing disease and controlling and improving nutrient availability.

TRANSGENICS

Although gene-based technologies have the potential to improve the efficiency of livestock production, thereby ensuring better returns for the farmers, the economic impact of transgenics in the livestock sector will be much less than in the crop sector.

However, the global adoption of genetically modified (GM) crops, which were grown on 67.7 million hectares in 2003 compared with 2.8 million hectares in 1996, has had a substantial impact on livestock feed. It is estimated that the United States of America (USA), Argentina, Canada, Brazil and the People's Republic of China have 63 per cent, 21 per cent, 6 per cent, 4 per cent and 4 per cent, respectively, of the global transgenic acreage and that the most frequently grown crops are GM soybean (61 per cent), maize (23 per cent), cotton (11 per cent) and canola (5 per cent).

Although few developing countries have released GM crop varieties, a preliminary analysis reveals that more than 20 developing countries are conducting research into the applications of GM crops. Although transgenic animals (especially mice) are used routinely in research (particularly in the medical field), no GM animals have yet been released on farms. A wide range of traits of potential interest to livestock producers have, however, been the subject of research; for example, the gene responsible for the production of growth hormone (which could be manipulated to increase growth rates), the phytase gene (which could reduce phosphorous emissions from pigs) and keratin genes (which could improve the wool of sheep).

The genetic modification of livestock has proceeded much more slowly than the genetic modification of crops for a variety of reasons, including the high costs, the inefficiency of the gene transfer techniques and the low reproductive rates of animals. Recombinant deoxyribonucleic acid (DNA) approaches have been used to promote the expression of desirable genes, to hinder the expression of undesirable genes, to alter specific genes and to inactivate genes so as to block specific pathways.

It is estimated that at least 30 enzymes produced by GM bacteria, yeasts and moulds are currently commercially available worldwide; many of these enzymes are used in the food industry. Genetic engineering has been used to introduce foreign genes into the animal genome or, alternatively, to knock out selected genes. Genes controlling growth were introduced into pigs to increase growth and improve carcass quality. Currently, research is underway to engineer resistance to diseases that affect the animals or that pose an indirect risk to human health, such as Marek's disease and salmonellosis in poultry, scrapie in sheep and mastitis in cattle.

Other studies have tried to increase the casein content of milk or to engineer animals that produce pharmaceutical or industrial chemicals in their milk or semen. No agricultural applications have yet proved commercially successful. Nuclear transfer (NT) technology now provides an alternative route for cell-based transgenesis in domestic species, offering new opportunities for genetic modification. Livestock that produce human therapeutic proteins in their milk, that have organs suitable for xenotransplantation and that are resistant to diseases such as spongiform encephalopathies have been produced by NT from engineered cultured somatic cells.

CHARACTERISING GENETIC VARIABILITY

There is considerable genetic diversity in the livestock of developing countries, much of which controls traits that influence adaptability to harsh environments, productivity and susceptibility to disease and parasitism. However, little if any data on these genetic resources are available. Economic analysis can play an important role in ensuring that conservation efforts are appropriately focused.

The primary challenge facing conservationists is to identify sound reasons why society should preserve animals that livestock keepers have abandoned. Jabbar and Diedhiou show that the breeding practices and breed preferences of livestock keepers can successfully be determined by using research techniques such as the revealed preference hedonic approach.

On the one hand, conservation cannot be achieved through a conventional breeding programme because the animals carrying the most advantageous traits cannot be easily identified; on the other hand, conservation cannot be achieved through biotechnology because the necessary technologies are either unavailable or uneconomic. In livestock populations with a high degree of genetic variation, molecular markers are being increasingly used to study the distribution and patterns of genetic diversity. Global surveys indicate that 40 per cent of domestic livestock breeds are at risk of extinction. Most of these breeds are found only in developing countries, and often little is known about them or their potential.

Rapid progress is being made in the preparation of dense microsatellite linkage maps to assist in the search for genetic traits of economic importance. These linkage maps can be used to develop strategies for marker assisted selection and marker assisted introgression that will meet the goals of breeding programmes in developing countries. Molecular markers have been widely used in the identification of genotypes and the 'genetic fingerprinting' of organisms. Genotype verification is used intensively to determine the parentage of domestic animals and to trace livestock products in the food chain back to the farm and animal of origin.

REPRODUCTIVE TECHNOLOGIES

The main objectives of using reproductive biotechnologies in livestock are to increase production, reproductive efficiency and rates of genetic improvement. Over the years, many options have become available for managing the reproduction of the major large and small ruminants. Artificial insemination (AI) and preservation of semen are the main technologies that are used extensively.

Assessing the fertilisation capacity of sperms, sexing sperms, synchronisation and fixed-time insemination, superovulation, embryo transfer (ET) and *in vitro* embryo production (IVEP) are additional techniques that can improve reproductive efficiency and pregnancy rates. Reproductive technologies can also be used to control reproductive diseases if procedures and protocols are accurately followed.

Artificial Insemination

The conception rate in field AI programmes in developing countries is very low, and therefore the desired effect in terms of animal improvement has not been achieved. Most semen banks still evaluate semen on the basis of sperm motility, even though significant advances have been made in techniques for semen evaluation.

Although detailed guidelines are available regarding the processing, storage and thawing of cattle semen and buffalo semen, the processing and handling procedures in laboratories processing semen are often inadequate. Only when farmers have access to considerably better technical and organisational facilities will AI become more effective. At present, the efficiency of the technology is limited by organisational and logistical constraints and by the failure to provide appropriate training for farmers.

Several modifications of the technique have been suggested to increase the conception rate. Synchronisation with different compounds, and the use of gonadotropin-releasing hormone (GnRH) followed seven days later by prostaglandin F2á (PGF2á) can synchronise oestrus and improves the conception rate.

In this protocol giving injections of GnRH on day 0, PGF2á on day 7 and GnRH on day 9 is called the 'Ovsynch' programme and synchronises ovulation, permitting timed insemination. The ability to control ovarian follicular and corpus luteum development has allowed insemination in cattle to be timed without having to detect oestrus, and this has increased the net revenue per cow.

Embryo Transfer

One of the major reproductive technologies that can facilitate genetic improvement in cattle is ET. Unfortunately, commercial ET programmes are limited by the high variability in the ovarian follicular response to gonadotropin stimulation. Multiple ovulation and embryo transfer (MOET) takes AI one step further, in terms of both the possible genetic gains and the level of technical expertise and organisation required.

In 2001, 450,000 embryos were transferred globally, mainly in dairy cattle, with 62 per cent being transferred in North America and Europe, 16 per cent in South America and 11 per cent in Asia. The main potential advantage of MOET for developing countries is that the elite females of local breeds can be identified, and bulls can be produced from them for use in a field programme of breed improvement.

Zebu cattle and buffaloes in developing countries exhibit less consistent follicular dynamics after superovulation than *Bos taurus* in the developed world. However, over the last 10 to 15 years, the number of transferable embryos produced by zebu donors has increased from 2.4 to 5.8 embryos per flush in the late 1980s to 5.6 to 9.9 embryos per flush in 2000. The use of ET has been less successful than envisaged for several reasons. The low reproductive efficiency, poor superovulatory responses, very low primordial follicle population and high incidence of atresia all contribute to low embryo production. In buffaloes, embryo recovery was initially less than one, but has subsequently improved to 2.6 with 1.4 transferable embryos per flush. After transferring buffalo embryos to recipients, the conception rate is only 16 per cent. The poor success rates have limited the use of ET in buffaloes, which are the main dairy animals in developing countries in Asia, South-East Asia and the Mediterranean region.

In Vitro Production of Embryos

Since the birth of the first buffalo calf from an *in vitro* fertilised oocyte, a number of publications have described the effects of different protocols and media on oocyte and embryo development. Two extensive reviews have been published recently. However, the practical use of IVEP is limited by high production costs and the low overall efficiency under field conditions. High rates of maturation (70 per cent to 90 per cent), fertilisation (60 per cent to 70 per cent) and cleavage (40 per cent to 50 per cent), and moderate to low rates of blastocyst formation (15 per cent to 30 per cent) and calf production (10.5 per cent) have been reported in the literature. The efficiency of blastocyst production in buffaloes is much poorer than the 30 per cent to 60 per cent reported for cattle.

Although viable buffalo blastocysts have been produced from ovaries obtained from abattoirs, the yield of transferable embryos remains low (15 per cent to 39 per cent). Embryos produced *in vitro* have led successfully to pregnancy and calf birth in buffalo, but the success rate is low. Therefore IVEP must be improved before it can be widely used in cattle and buffaloes in developing countries.

9

Farming Tools

GPS: ITS USES AND POTENTIAL ARE GROWING

Global positioning systems (GPS) are widely available in the agricultural community.

Farm uses include:

- Mapping yields
- Variable rate planting,
- Variable rate lime and fertilizer application,
- Field mapping for records and insurance purposes, and
- Parallel swathing.

GPS and associated navigation systems are used in many types of agricultural operations. These systems are useful particularly in applying pesticides, lime, and fertilizers and in tracking wide planters/drills or large grain-harvesting platforms. GPS navigation tools can replace foam for sprayers and planter/drill-disk markers for making parallel swaths across a field.

Navigation systems help operators reduce skips and overlaps, especially when using methods that rely on visual estimation of swath distance and/or counting rows. This technology reduces the chance of misapplication of agrochemicals and has the potential to safeguard water quality. Also, GPS navigation can be used to keep implements in the same traffic pattern year-to-year, thus minimizing adverse effects of implement traffic.

Use of GPS navigation in agrochemical application with ground equipment has grown rapidly, and commercial applicators are quickly adopting the tool. A 2007 survey of those who offered custom application, 82 per cent applied at least some of the fertilizer/chemicals using a GPS navigation system with a manual-control/lightbar guidance system.

Twenty-nine per cent said they used a GPS navigation system with an auto-control/auto-steer guidance system for at least some of their custom application. On average, for all custom applied materials, 57 per cent was applied with GPS lightbar, and 12 per cent was applied with auto-steer GPS. GPS navigation has become standard practice for US aerial applicators. Crop producers are also starting to adopt these systems, because GPS navigation is an excellent way to improve accuracy, speed, and uniformity of application.

NAVIGATION SYSTEMS IMPORTANT TO FIELD OPERATIONS

Automated guidance of agricultural vehicles has been motivated by a number of factors—most important is to relieve the operator from continuously making steering adjustments while striving to maintain field equipment or implement performance at an acceptable level. This is not surprising, considering the many functions an operator must monitor, perform, and/or control while operating the vehicle. The requirements placed on farm-equipment operators have changed drastically with increases in equipment size, power, multiple equipment functions, and speed—as well as monitors reporting on specific system performance. These increasing demands on the operator can result in increased errors in function, costs, environmental problems, and operator fatigue.

Foam Markers, a Widely Used Technology

Foam markers are the most common form of navigation aid used during fertilizer and pesticide application. The foam is dropped and used to align the applicator dur-ing the return pass. Foam markers utilize an air pump to pressurize a tank containing the foaming agent. The pressurized fluid causes the foaming agent to flow into an accumulating chamber. The foam collects in this chamber until the accumulated mass overcomes surface tension, causing a foam blob to fall to the ground.

Most often the foam accumulators are placed at the ends of the applicator boom or, alternately, at the centre of the applicator when booms were not utilized, as in the case with spinner-disk granular applicators. Equipment operators use the foam blobs left on the field surface as a navigation aid to know where the applicator has passed.

GPS + NAVIGATION AIDS

Relatively inexpensive navigation aids known as parallel-tracking devices assist the operators to visualize their position with respect to previous passes and to recognize the need for steering adjustments. These aids are commercially available in several configura-tions. One system is a lightbar, which consists of a horizontal series of *Light Emitting Diodes* (LEDs) in a plastic case 12 inches to 18 inches long.

This system is linked to a GPS receiver and a micro-processor. The lightbar is usually positioned in front of the operator, so he or she can see the accuracy indicator display without taking their eyes off the field. The lightbar can be mounted inside or outside of the cab, and the operator watches the "bar of light". If the light is on the centerline, the machine is on target. If a bar of light extends to the left, the machine is off the path to the left and needs to be corrected. In like manner, if a bar of light extends to the right, the machine is off to the right.

Software allows the operator to specify the sensitivity to and distance between the swaths. Similar GPS navigation systems have been used for aerial application since the early 1990s. Also, the GPS system gives the current location of the implement and, with past traffic patterns, the computer interface provides the operator directions to maintain proper swath width to match adjacent traffic paths.

If an operator leaves the field to refill the applicator or is forced out of the field due to weather, upon return, the operator can resume and maintain accurate swath widths, and over-spraying on previous sprayed areas is eliminated. More advanced systems have a screen showing the swath of the machine as it moves through the field. Early models only allowed straight-line parallel swaths, but current models are available for any contour traffic pattern. Areas covered with previous swaths are indicated on the screen. The advanced navigation system coupled with a variable rate spreader drive and software has the capacity to generate "as-applied" maps showing previous coverage and the application pattern.

This provides an excellent record of the pattern and timing while operating in the field. Portions of the field that are not treated, such as wet areas, can be marked in the computer and stored for later operations when conditions permit application. All of this is done without having to physically mark the field area with flags.

GPS + AUTO-STEER NAVIGATION

More advanced navigation systems possess similar capabilities as the navigation aids and also have the additional option to automatically steer the vehicle. Auto-steering is accomplished with a device mounted to the steering column or through the electro-hydraulic steering system. The accuracy level of these systems is based on the quality of differential correction and internal data processing: as the accuracy improves, the corresponding cost increases.

These navigation systems are classified in three categories:

- *Submeter Accuracy:* Submeter accuracy usually means approximately two feet to four feet year-to-year, and less than one foot pass-to-pass errors. The differential correction source could be from a Coast Guard beacon, WAAS (Wide Area Augmentation System), or satellite providers. These systems are relatively inexpensive and can be used while performing tillage, some types of fertilizer and chemical applications, seeding, and harvesting. These devices can be easily transferred between vehicles, so the same steering system can be used on different vehicles. However, operations requiring highly accurate guidance are not feasible with submeter level equipment.

- *Decimeter Accuracy:* Decimeter accuracy provides approximately four inches to eight inches year-to-year and three inches to five inches pass-to-pass errors. This can be achieved using either a local base station or dual frequency receivers with private satellite differential correction subscription. With the increased performance, operators can use auto-steering during most of the conventional field practices. Prices range from $15,000 to $25,000 plus the satellite subscription.
- *Centimetre Accuracy:* Centimetre accuracy can be obtained by using a local base station with the *real-time kinematic* (RTK) differential correction. Both long-term and short-term errors have been reported for these systems. Vehicles equipped with this high-level equipment can be used to conduct strip tilling, drip-tape placement, land leveling and other operations requiring superior performance—as well as virtually any other task. In addition to the ability to accurately determine geographic location, auto-guidance systems usually measure vehicle orientation in space and compensate for unusual attitude, including roll, pitch, and yaw. The price ranges from $40,000 to $50,000 with no annual subscription fees.

GPS Navigation vs. Foam Markers

Potential advantages of GPS Navigation relative to foam markers for agricultural applications include:

- *GN is more Reliable and more Accurate than Foam Markers*: Using foam markers could cause about 10 per cent of the field to be skipped or overlapped. With the GN, the skip and overlap rate drops to about 5 per cent. Some tests have shown that with an experienced operator, the skip and overlap rate can be as low as 1.5 per cent.
- *GN Allows Accuracy at Higher Speeds:* GPS navigation can attain a 13 per cent to 20 per cent higher speed than a foam marker. Naturally, an increase of speed is terrain dependent. If field conditions limit speed, then a GN benefit is unlikely.
- *GN is a Possibility with Spinner Spreaders*: Foam markers are not generally used with spinner spreaders. The spreaders have no boom on which the foam equipment can be installed. Due to the spread width, a foam marker in the centre of the machine path is difficult

to see from the next swath, and the driver would still be using a visual estimate for the spreader swath.

- *GN is Easy to Use*: Anybody can learn to use GPS navigation systems, regardless of computer skills. The systems require only a little practice—typically about 30 minutes.
- *GN Provides Effective Guidance over Growing Crops:* With solid-seeded crops, foam tends to fall through the canopy to the ground where it is almost invisible, contributing to skipping or overlapping. Crop height does not affect GPS.
- *GN Allows Operation when Visibility is Poor:* GN works at night, in dust, or in fog. This lengthens working time during critical planting and spraying periods. In many areas, nighttime is best for spraying because of low wind speed. Leaf orientation changes at night and herbicide may or may not be as effective as during daylight. Nighttime spraying of insecticides can protect sensitive bee colonies.
- *GN is Less Affected by Weather*: In certain conditions such as low humidity, heat, and large fields, foam markers can be rendered ineffective. The foam can evaporate before the operator makes the return swath. GN works at any temperature, including low temperatures when foam systems freeze.
- *GN has Lower Recurring Costs:* GN has no moving parts or tubes to clog. Depending on the manufacturer, software updates are usually free to system owners. Foammarker systems require foam, dye, and tank cleaner.

The primary recurring cost for GPS navigation is satellite differential correction. Typically, the subscription costs $600 to $1,500 per year for each GPS unit. Many producers already have GPS for yield monitoring and may subscribe for differential correction. For them, GN has almost no recurring costs. In many areas of Virginia, it is possible to use the Coast Guard beacon or WAAS for GPS differential correction. These services are free and may be adequate for some applications, including dry fertilizer with a spinner spreader.

Accuracy for spraying should be within six inches; the GPS equipment supplier can pro-vide information regarding the differential correction needs for a specific location:

- *GN Reduces Operator Fatigue and Eye Strain*: With the lightbar mounted directly in front, GN operators do not need to look backward or sideways. They can drive accurate swaths while looking straight ahead.
- *GN has Lower Setup Time:* Foam markers require that tanks be filled

and dyes be changed. GN begins working approximately 30 seconds after the machine is switched on.

- *GN is not Affected by Wind or Boom Bounce*: Blowing foam or a foam system bouncing over rough ground at the end of a long boom may intro-duce substantial error. GPS systems are not affected by rough terrain or wind.
- *GN Reduces Pesticide Use by Reducing Overlaps*: If a 10 per cent overlap is reduced to 5 per cent, pes-ticide use also is reduced by 5 per cent. The same is true for fertilizer and seed, so using GN is better for the environment and good for the bottom line.
- *GN Reduces the Need for People to Enter Already Sprayed Areas*: The operator can mark where spraying stopped without dismounting.

Additional Features

Decisions about adapting GPS navigation should be based on particular farm needs, farm operation procedures, and understanding of positioning errors. Such issues as vehicle dynamics, tracking of trailed implements, and field terrain also need to be considered. Proper alignment and installation of the GPS navigation system is required for effective field operation. Poor quality of the steering-control systems, a sloped terrain, or misaligned implements will cause the field performance of GPS navigation to suffer.

An important feature of GPS navigation systems is the ability to follow particular traffic patterns and provide effective feedback so that the operator or auto-steer system can appropriately respond. Most systems can effectively perform straight-line patterns and many can follow contours and other field features. Although most GPS navigation systems are designed specifically for the task of vehicle steering, some systems have features to collect spatial data or to operate variable-rate controllers. Additional features spread the cost of the system among several tasks while increasing efficiency in multiple areas of the crop production programme.

The documentation provided by a GPS system could be used to show that the product or field work was applied in specific locations at a specific time. Setbacks specified by regulatory agencies and other sensitive areas may be defined as "no-apply zones". The "as-applied" map includes areas in which application occurred and rates can be recorded. It is possible to include not only position, but time and date information in an application record as well. Such documentation could be used for accurate record–keeping for agrochemical application and machine operation and, potentially, for regulatory and legal defence purposes.

GPS navigation products are distinguished by the compactness of their components and user interfaces. Some systems cause technical challenges during installation and calibration, while others may be functional in less than an hour. User interfaces range from intuitive, colorful, graphic touch-screen displays to hard-key menus with limited graphics feedback.

A sloped terrain makes control of vehicle dynamics challenging. Roll, pitch, and yaw alter the GPS antenna location with the projected centre of the vehicle. For example, when driving across a slope, the horizontal position of the GPS antenna is downhill with respect of the centre of the vehicle, and the guidance will be in error down the slope. To compensate, some systems include gyroscopes, accelerometers, or additional GPS antennas. Less advanced terrain-compensation modules deal with only roll and pitch angles of the vehicle, while others can measure total dynamic attitude in six degrees of freedom and enable the system to compensate for variable terrain.

Several manufacturers have add-on implement steering systems for proper implement tracking. Compensating for known geometry of the tracking from implements can be adjusted with the vehicle's trajectory to keep the implement aligned. Optical and crop-based tracking systems can be useful when positioning implements with respect to established rows.

GPS Navigation Cost

The most frequently mentioned disadvantage of GPS navigation is the up-front cost. Costs range from about $1,500 for a farmer who already owns a GPS to more than $14,500 for a commercial applicator who wants a system that keeps "as-applied" records as well as pro-vides navigation. A fully automatic navigation system that steers a tractor, sprayer, or combine—with operator engagement only at field ends—could range from $6,000 to $50,000.

Current swathing aids such as foam marker systems range from $500 to $3,000. Speed is also an issue in foam systems. The lower-cost foam systems are slower, but they work adequately when application is done with a tractor. Commercial applicators operating at 20 miles per hour need more foam output than some systems can provide.

Costs and benefits vary widely depending on the crop, acreage covered, swathing accuracy achieved, and other factors. As a rule-of-thumb, a navigation system could cost six times more than a foam marker system, which means that justification for GPS navigation over foam markers must be computed from the benefit side

Estimates of the economic impact of sprayer skips are complicated because influence of weed control on crop yield varies by crop. For example, the weed population and long-term weed seed-bank effects have to be evaluated and assessed. For fertilizer and lime application, how much yield is lost if the area receives no application? A skip is much more costly in a higher-value crop, such as vegetables or seed crops, than it would be in a bulk commodity such as corn, soybeans, or wheat.

If the skip occurs in a very clean field, the yield-loss effect due to reduced weed control may be minimal, but in a heavily infested field, the yield may drop to almost zero in the skipped area. Weed scientists suggest that the greatest economic effect of skips may be the creation of a seed bank that will lead to management problems and greatly increased weed control costs in future years. Similarly, for a field at pH 5.8, the yield loss due to skipping an area with lime application will probably be small during the first year, but will become greater in later years.

Bottom Line

GPS navigation has many advantages over conventional marking devices such as foam markers, and especially over the visual-estimation method for spinner spreaders. With an existing GPS being used for yield monitoring or field mapping and soil sampling, the GPS navigation system can increase the efficiency of the farm or agribusiness while minimizing adverse environmental impacts associated with overlapping applications.

The system can also reduce operator fatigue and anxiety regarding fertilizer and pesticide application. Finally, use of this technology can demon-strate to the non-agricultural community that advanced technology is being used to farm efficiently and safely. The advantage of "as-applied" maps, provided by some systems, is documentation that applications were made at the appropriate location and rate.

The history of agriculture dates back thousands of years, and its development has been driven and defined by greatly different climates, cultures, and technologies. However, all farming generally relies on techniques to expand and maintain the lands suitable for raising domesticated species. For plants, this usually requires some form of irrigation, although there are methods of dryland farming; pastoral herding on rangeland is still the most common means of raising livestock. In the developed world, industrial agriculture based on large-scale monoculture has become the dominant system of modern farming, although there is growing support for sustainable agriculture.

Modern agronomy, plant breeding, pesticides and fertilizers, and technological improvements have sharply increased yields from cultivation, but at the same time have caused widespread ecological damage and negative human health effects. Selective breeding and modern practices in animal husbandry such as intensive pig farming have similarly increased the output of meat, but have raised concerns about animal cruelty and the health effects of the antibiotics, growth hormones, and other chemicals commonly used in industrial meat production.

FARMING SYSTEMS

The availability of food has always been the preoccupation of mankind. Despite doubling of the global population in the past four decades, farmers have produced sufficient food to allow average per capita food intake to grow gradually. Yet hunger persists and food reserves have fluctuated market during this period; sometimes falling below critical low levels. Every person has a right to have access to safe and nutritious food and be free from hunger according to the commitment of the world food summit of FAO 1996. To reduce the number of the undernourished people, eradication of poverty and hunger must be tackled.

Farming systems analysis will recognize the diversity of livelihoods of poor farmers, pastoralists and fishing families and explore various pathways that may help them escape from poverty and hunger. The analysis may also give directions of change in the major farming systems throughout the developing world for the future. Rural development depends on outcome of daily decisions of farmers. The challenge to governments, civil society organizations and private sector is to provide public goods, institutional environment and incentives to enable farm households to enhance agricultural growth and poverty reduction. Therefore, the policies and technologies have the key to strategic opportunities for action for the different farming systems, for each developing region and for the developing world as a whole.

The opportunities should focus on closing gaps between the projected slow reduction of hunger and poverty and the goals set by the international community in the millennium declaration. The challenge for developing states is to identify specific agricultural and rural development needs and opportunities and focus investment in areas where food insecurity and poverty will be reduced. This identification and resource allocation process can be facilitated by analysing farming systems in order to understand the local factors and linkages. This will form the development constraints and investment opportunities in a farming systems framework.

CATEGORIES OF FARMING SYSTEMS

Delineation of major farming systems provides a useful framework within which appropriate agricultural develop-ment strategies and interventions can be determined. To avoid the complexity of discrete, micro-level farming systems in each developing country, only major farming systems have to be identified and mapped to estimate the magnitude of their populations and resource bases.

EVOLUTION OF FARMING SYSTEMS

The farming systems approach considers both biophysical dimemensions and socio-economic aspects at the level of the farm-where most agricultural production and consumption decision are made. The power of the approach lies in its ability to integrate multidisciplinary analyses of production and its relationship to the key biophysical and socioeconomic determinants of a farming system. In order to present the analysis of farming systems and their future development within a framework that is broadly comparable between systems and across different regions, the biophysical and socio-economic determinants are grouped into five categories:

- Natural resource and climate;
- Science and technology;
- Trade liberalization and market development;
- Policies, institutions and public goods; and
- Information and human capital

These categories represent the major areas where the major farming systems characteristics, performance and evolution are likely to be affected over the next 30 years. The interrelationship of the key determinants of farm systems and by extension farming systems. Some of these factors are internal to, or part of, the farming system, others are external.

THE DRIVES INFLUENCING FARMING SYSTEMS EVOLUTION

NATURAL RESOURCES AND CLIMATE

The interaction of natural resources, climate and population determine the physical basis for farming systems. In the early stages of development, population increase led to expansion of cultivated area. These create conflict of different users of land and water resources. Once good arable land is exploited, further population increase leads to intensification of farming systems. As forests and woodland come under pressure, biodiversity is threatened hence tension between development and conservation goals.

Water Availability

As development of irrigation systems increase, about 70 per cent of total volume of water is used. There is going to be a limit to the use of this water, hence reduce current rates of irrigation growth. The expansion of agriculture, plus changes in production technology has resulted in a decrease in agro-biodiversity. This is reduction in indigenous flora and fauna, varieties cultivated especially maize, wheat and rice and loss of biodiversity in domestic animals. The remedy could be in modern biotechnology.

Climate

Agriculture contributes about 30 per cent anthropogenic emission of green house gases:

- Methane production by livestock is expected to increase. This is an indication that global climate change will be significant. Average surface temperature globally is expected to rise by an estimated 1.4 to 5.8°C in the next 100 years

- Expected also is a dramatic increase in climatic extremes. High temperature will lead to rise in sea levels–estimated at 0.1 and 0.9 metres over this century. Therefore not doubt that agriculture and food security will be affected by climate change. Other threats include diseases, and hurricanes.

Science and Technology

For the last four decades, there has been a rapid expansion in investments in agricultural sciences and technology. In the 70s, the Consultative Group on International Agriculture Research (CGIAR) and national agricultural research systems were strengthened.

These have improved production technologies with emphasis on improved yielding varieties. FAO projections to 2030 indicate a continuing rise in average cereal yields in developing states under both rain fed and irrigated systems: However, the small-holder poor farmers in marginal areas have not benefited from cereal yield increases and investments in technology development in non-cereal crops. Investment in livestock research has not been commensurate with the contribution of the sub-sector to household income. Only one CGIAR research centre, the International Livestock Research Institute (ILRI) in Kenya concentrates on livestock research.

There has been far less research in integrated technologies for diversifying the livelihoods of small-holder farmers in developing states and increasing the sustainability of land use. For instance, the role of organic matter in soils, development of reduced tillage systems, use of on-farm organic resources in combination with inorganic fertilizers, role of legumes in biological nitrogen fixation and Integrated Pest Management (IPM) and weed and pest control. These may be of little interest to the private sector and can be neglected by the public research institutions.

However, global research agenda focuses on moving away from individual crop performance to increased system productivity. This is viewed in terms of better managed interactions among farm enterprises, sustainable resource management and improved targeting of technologies towards women farmers and poor households.

Institutional modalities are changing from public sector focus to public–private partnership driven by client demand due to growing under-standing of farmers' problems and opportunities and the willingness of blending indigenous knowledge and modern information. Biotechnology is focusing on profit–generating inputs, export crops and agro-processing.

Trade Liberalization and Market Development

This is mainly on economic reform and trade liberalization. By end of 70s, economies of many developing states had been distorted as a result of excessive government intervention and control.

The economies were characterized by:

- High inflammation,
- Unmanageable balance of payments and fiscal deficits,
- High external debt ratios and
- Gross domestic product (GDP) growth rates that were negative or failing to march the rate of population increase.

To solve these problems, the international monetary fund (IMF) and World Bank and other international institutional donors initiated lending programmes with a condition of adoption of programmes of structural reform–best known as Structural Adjustment Programmes (SAPs).The SAPs have developed into liberalized trade specifically for agriculture sector.

For example measures to:

- End marketing monopolies,
- Reduce parastatal involvement in supply of inputs, marketing and processing,
- Reduce or remove subsidies, price controls and impediments to private sector activities,
- Remove restraints on foreign trade, and
- Promote the private sector.

Small-scale activities requiring limited management, technical knowledge and with limited capital requirements have been rapidly adopted. The first major agricultural service to be privatized in many developing states was grain marketing and milling. This was because grain marketing boards had imposed major financial burdens on governments. International agreements establishing World Trade Organizations (WTO) has boosted trade liberalization.

Markets have a great role to play in agricultural development as they form linkages between farm, rural and urban economies that the development process depend. As a result of these changes, there is change in structure of production at all levels including; small-holder farming systems in patterns of production and natural resource usage in response to market forces. The speed of change has had negative impacts. Poverty increased in many farming systems as a result of reduction in government support and declining prices for major small-holder products.

Policies, Institutions and Public Goods

The development of farming systems requires a conducive policy environment. Establishing farm-rural-urban linkages requires effective demand. The greatest change in the policy environment in the past 30 years has been; structural adjust-ment and the decline of national food self-sufficiency as a major element in shaping of policies for rural areas. In the 60s, the need to ensure national food security was paramount for many governments and this justified direct intervention in agricultural marketing, storage, import licensing and import subsidies, among others.

National food security is no longer an overriding policy aim; food security remains key policy issue for developing states and the world. As SAPs progressed, policymakers shifted their attention to the potential to increase the efficiency of service delivery through restructuring of institutions. These include: shift of traditionally public roles to civil society and private sector, decentralization of government services and reduction of government investment in provision of public services. This approach will encourage more local participation in decision making and resource allocation.

While such trends have offered significant benefits in mobilizing of non-governmental resources and aligning of public activities to local needs, they have also created certain constraints. Such constraints include; slow or erratic supply response in finance, research, extension, education, health and infrastructure in rural areas where poverty is widespread. There is a feeling of creating new public services that can support the environment for the growth of private sector activities and to ensure equity and environmental sustainability.

This could strengthen women participation in local governance, introduce individual household incentive policies, have access to, and control of natural resources. This will create intensification and hence call for proper land policy and water management.

CHALLENGES OF INDIAN AGRICULTURE

Indian agriculture is facing a policy paradox. Although several forecasts of the 1990s predicted that India would be a large importer of grains in the years to follow, in fact from 2001 to 2004 India exported around 30 million tons of foodgrains. It was seeking primarily to liquidate its bulging grain stocks, which reached 63 million tons in July 2002. Whereas India's agricultural policy is still rooted in the goal of self-sufficiency in grains, consumption patterns are changing fast towards high-value agricultural products such as fruits and vegetables, livestock products, and fish.

The policy environment is lagging behind the structural change occurring in India's consumption and production baskets. On another front, foreign exchange reserves, which had reached a rock-bottom US $1.2 billion in July 1991, climbed to more than US $120 billion by the end of 2004.

Nonetheless, despite comfortable food and foreign exchange reserves and reasonably high growth in Gross Domestic Product (GDP) of about 6 per cent annually, India still has more than 250 million underfed people and high under-employment.

This situation reflects severe problems on the distribution front. What are the reasons behind this paradoxical situation? The answer presumably lies in the neglect of, as well as misallocation of resources in, agriculture and rural develop-ment, especially in the later phase of the reform process initiated in 1991. The average annual rate of growth in agriculture fell from more than 4 per cent per year during 1992/93 to 1996/97 to less than 2 per cent per year during the period 1997/98 to 2002/03, and it remains low.

What led to this dramatic decline in the growth of agriculture since 1997/98? How can it be revived?

How can growth in agriculture and rural development diminish poverty quickly? To stimulate pro-poor agricultural growth and rural development, India will need to make some strategic choices. We propose action in five major areas that can help the government to accelerate agricultural growth and reduce poverty, malnutrition, and unemployment quickly and on a sustainable basis. All of these reforms can be achieved with due regard for the well-being of the country's rural poor.

AGRICULTURAL CHANGE IN PLANTINGS

Agricultural change refers to the difference between the first plantings 10,000 years ago and today's computerized, industrialized, genetically engineered production systems; but a process that occurs on a daily basis, as farmers make decisions about what, where, and how to cultivate. It goes well beyond how much food is produced, how much money is made, and how the environment is affected: Agriculture is intimately linked to many institutions in every society, and to population.

Agricultural change is mainly based on sustainable livelihoods and agricultural intensification. A livelihood is sustainable if it can cope with, and recover from, stresses and shocks, maintain or enhance its capabilities and assets and provide net benefits to other livelihoods locally and more widely, both now and in the future, without undermining the natural resource base. Agricultural intensification is increased average inputs of labour or capital on a small–holding, either cultivated land alone, or on cultivated and grazing land, for the purpose of increasing the value of output per hectare.

It may occur as a result of an increase in the gross output in fixed proportions due to inputs expanding proportionately, without technological changes, a shift towards more valuable outputs or technical progress that raises land productivity. However a shift from a high yielding crop of 100 days per year to two low-yielding crops of 250 days per year cannot be described as intensification, though it may shift output value towards labour.

The international agricultural development policy based on CGIAR denies developing world small-holders and nations an opportunity to diversify food production, adapt new improved local techniques of production and resist global market competition. Through this international concept of development, techniques and resources of the developing world are devalued. An increase in living standards is hinged on the production of cheap food and non-food agricultural products, for global market competition at the expense of social and ecologic characteristics. Globalization has come at the price of undermining some of these characteristics like diversity, adaptability, and resilience in exchange for the overreaching goal of efficiency.

Meanwhile, citizens of these externally dominated nations starve as tons of flowers are delivered to other states from their soil. Unless one adopts a coldly Malthusian perspective, the situation is worthy of examination and further study. Aside from the degradation of human life imposed by economic agricultural development policy, there are informational costs. The loss of diversity, biotic and cultural, further degrades the quality and quantity of information available regarding natural farming techniques.

As indigenous farming knowledge gives way to chemical sprays and imported seeds, not only the perceived, but actual value of that knowledge is diminished. Onfarm recycling is necessary for locally sustainable farming and contradicts the conventional high-external input approach. Local food markets provide a means to confine inputs and outputs in the current context of needs and sustainability of the community. In order for these outside forces to work in balance with the farm, there must be shared goals. The conflict between the state and the individual arising within the modern agricultural paradigm is the result of conflicting agendas between the top and the bottom.

CHANGE PROCESS AFFECTING AGRICULTURAL PRODUCTION AND LIVELIHOODS

The lack of transparent, timely and reliable crop and livestock marketing information is seen by many as one of the greatest challenges to the development of the agriculture industry in East India. Without access to information about agricultural products prices at different regional markets, farmers are unable to identify which points-of-sale offer the best prices for their agri-products. For example, in Mbirikani, Kajiado County, Kenya, Maasai pastoralists once relied on middlemen to gain market information. However, a marketing information system has been developed by the Livestock Information Network and Knowledge System (LINKS) to support livestock producers and traders. Now the farmers use mobile phones to ask for livestock prices in Emali, Mombasa and Nairobi, and this knowledge significantly improves their bargaining power with livestock traders.

INCREASING MARKETING OPPORTUNITIES

Given the high dependency of pastoralists' livelihoods on the sale of livestock and livestock products, LINKS designed and implemented an Information Communication Technology (ICT) infrastructure to collate data on livestock sales and prices from a network of different markets for dissemination using SMS messages. This has been a success and has been used to develop a National Livestock Marketing Information System (NLMIS) in Kenya. NLMIS relies on a network of County livestock marketing officers and data monitors to transmit data between markets and the database.

The field officers are also trained to download, analyse, and summarize the information to transmit to pastoralists and traders, including information about prices and volumes of cattle, camels, sheep and goats at livestock markets. The market information is also made available online, and may be downloaded for printing and sharing with livestock communities with no internet access. The information is also shared by e-mail, posted on billboards at market places and can be requested through SMS.

Vibrant Markets

This change in communication has also transformed undeveloped markets in other value chains in agriculture. The availability of marketing information through SMS and information boards, competition among agri-products traders has increased, and this has improved and stabilized prices paid to farmers. Women have also benefited by providing animal health services, fodder and small loans to the traders.

CHALLENGES IN PRODUCTION FOR RURAL LIVELIHOODS AND OPPORTUNITIES

Some of the challenges that affect production in the rural livelihoods in developing states, especially in India include:

CLIMATE CHANGE

Climate change is exacerbated by other factors like widespread poverty, recurrent droughts and floods. Others are dependence on natural resources and biodiversity, over–dependence on rain-fed agriculture, heavy disease burden, and numerous conflicts that have engulfed the continent. Changes in future climate may affect negatively the overall economy of India, hence hampering its potential economic growth.

Bio-fuel

From living organisms or the waste these are produced, especially from crops such as maize, soya, sugarcane and rapeseed.

These bio-fuel are of two types:

- Bioethanol made from sugarcane, sugar beet and cereal crops, and
- Bio-diesel made from soybean, rapeseed, vegetable oils, animal fats used as frying oils.

The question being asked is:

- Is bio-fuel production a solution to climate change or cause of global hunger? If made from plants, is it environmental friendly as opposed to fossil fuels. It is argued that crops absorb carbon dioxide from the atmosphere as they grow and produce about 60 per cent less carbon dioxide than fossils. But land previously used for food crops is being turned over to bio-fuels.

POPULATION AND MIGRATION

Population growth is an important factor for local environment change. Rapid population growth is one of the main reasons for increasing the number of people. The move for livelihood and migration has been a major factor of rapid population growth in urban areas in less developed states.

Urbanization

It is the driving force for modernization, economic growth and development. However, there is increasing concern about the effects of expanding cities on human health, livelihoods and the environment. The implications of rapid urbanization and demographic trends for employment, food security, water supply, shelter and sanitation, especially the disposal of wastes that the cities produce are staggering.

The question that arises is whether the current trend in urban growth is sustainable considering the accompanying urban challenges such as unemployment, slum development, poverty and environmental degradation, especially in the developing states.

ROLE OF AGRICULTURAL INFORMATION

Although agriculture is the backbone of the economies of most Indian states, development in this sector has not grown as fast as the population. Women in India are the key actors in agriculture; and produce about 80 per cent of the region's food.

The major problems include cultural, social, economic, legal, education and lack of information to improve farming activities. Lack of reliable and comprehensive information is one of the major hindrances to agricultural development.

Population growth posses the following challenges in production and livelihood opportunities: Intensification of agricultural production, soil degradation, *e.g.* erosion, lack of access to land, reduced livestock grazing possibilities, and limits to grazing land. Others challenges include degradation of water resources, difficult access to irrigation water, decreased size of food portions and declining nutrition, deteriorating health standards, problems accessing credit, and difficulty in accessing education and training.

Still others include demographic trends towards a younger population places increasing pressure on labour markets, resulting in greater unemployment and out-migration, increased political instability and conflicts, increases affluence gap between rich and poor regions which has implications for migration, and it fuels the desire to emigrate from poor regions. Yet there is increased potential for demographically fuelled conflict resulting from declining livelihood assets.

Further, environmental problems include destruction of natural ecosystems, increased rate of species extinction, falling water tables and depletion of aquifers, pollution of rivers, seas and coastal waters, and increase of harmful emissions to the atmosphere.

CHALLENGE TO PRODUCTION AND RURAL LIVELIHOOD OPPORTUNITIES

Men and women suffer similar problems, but these seem to affect women more adversely. Unfortunately, access to information has not received adequate attention in most states and especially in rural aresas where 70–80 per cent of the Indian population live. Women in rural areas have very little access to information, and most are poor and cannot read or write. Rural and peri-urban women, therefore, need to be empowered to increase agricultural productivity through access to information.

The few assessments that have been conducted on the information needs of women in agriculture have shown that women require information on all aspects of agricultural production, processing, marketing, decision-making processes, the resource base, and trade laws. They also need to exchange information on indigenous knowledge and require appropriate Information Communication Technologies (ICTs) to be able to access vital information efficiently and cost-effectively.

THE ROLE OF ICT AS A COMMUNICATION TOOL

Traditional ICTs such as drama, dance, folklore, group discussions, meetings, exhibitions, demonstrations, visits, farmer field schools, agricultural shows, radio, television, video and print media have been used successfully in many Indian states. The media play a major role in delivering agricultural messages, with the radio forming the main source of information. Satellite, solar and fibre optic technologies are now in use for computers, telephones and facsimile. Where appropriate, these should be tapped to enable rural women farmers to access information using modern ICTs concurrently with traditional ones.

Telecentres are the way forward and could be established in villages and new ICTs such as electronic mail, World Wide Web, electronic networks, Newsgroups, ListServs, telecon-ferencing, CD-ROM and distance learning tools can be used where appropriate by rural farmers, farmer leaders or their intermediaries such as non-governmental organizations, community based organizations and development agencies. These intermediaries would in turn repackage the information for the consumption of rural farmers including women.

The agricultural information sector has a relatively weak technological base and insufficient scientific, technical and educational capacity and thus needs substantial capacity building. There is also a dire need for radical change in agricultural information systems, if widespread management and dissemination of information and knowledge in digital format is to take root. This change can be driven by building partnerships among agricultural information professionals in the region, the government and international and regional partners such as FAO, AARINENA/RAIS, ICARDA, AOAD and ASARECA.

PRODUCTIVE CAPACITY AND CHANGE OF AGRO-ECOSYSTEMS AS A CHALLENGE TO LIVELIHOOD OPPORTUNITIES

Agro–ecology is a scientific discipline that uses ecological theory to study, design, manage and evaluate agricultural systems that are productive but also resource conserving. It considers interactions of all important biophysical, technical and socio–economic components of farming systems. Broadly stated, it is the study of the role of agriculture in the world. Agro–ecology also provides an interdisciplinary framework with which to study the activity of agriculture.

In this framework, agriculture does not exist as an isolated entity, but as part of ecology of contexts. Agro–ecology draws upon basic ecological principles for its conceptual framework. It refers to an ecological approach to agriculture that views agricultural areas as ecosystems and is concerned with the ecological impact of agricultural practices. Agro–ecosystems on the other hand, refer to a model for the functioning of an agricultural system, with all inputs and outputs.

An ecosystem may be as small as a set of microbial interactions that take place on the surface of roots, or as large as the globe. An agro ecosystem may be at the level of the individual plant-soil-microorganism system, at the level of crops or herds of domesticated animals, at the level of farms or agricultural landscapes, or at the level of entire agricultural economies. They analyse; mineral cycles, energy transformations, biological processes and socioeconomic relationships as a whole in an interdisciplinary fashion.

Further, they are concerned with the maintenance of a productive agriculture that sustains yields and optimizes the use of local resources while minimizing the negative environment and socio-economic impacts of modern technologies. In industrialized states, modern agriculture with its highinput technologies generates environmental and health problems that often do not serve the needs of producers and consumers. In developing states, addition to promoting environmental degradation, modern agricultural technologies have bypassed the circumstances and socio-economic needs of large numbers of resource-poor farmers.

Agro-ecology is the use of ecological theory to evaluate agricultural systems that are productive and resource conserving. It encompasses interactions of biophysical, technological and socio-economic components of farming systems. It is characterized by energy and nutrient flow, biodiversity and evolution. The main challenge to agroeco-systems is the modern technology and may also include social, cultural and economic aspects of life. The main drivers of the agro-ecosystems change are agricultural expansion, intensification of production and environmental concerns including climate change.

Policy issues are of paramount importance. Therefore, Ecological agriculture aims at improving agricultural production and post-production while conserving the regenerative and reproductive capacity of the natural resource base, and thus, avoiding the cycle of rectification of error. It is based on traditional ecological understanding and combines results of modern science of natural processes. Hence, ecological agriculture is about maximizing use of knowledge of natural processes. Criteria for ecological agriculture need to consider technical performance, efficiency, impact, resource use, availability, and user preference.

CLIMATE CHANGE AND SHIFT TO BIO-FUELS: A CHALLENGE TOPRODUCTION AND LIVELIHOODS

Livelihood comprises of the capabilities, and material and social assets necessary for a means of living. On the other hand, Climate change is the natural cycle through which the earth and its atmosphere are going to accommodate the change in the amount of energy received from the sun. The climate goes through warm and cold periods, taking hundreds of years to complete one cycle. Changes in temperature also influence the rainfall, but the biosphere is able to adapt to a changing climate if these changes take place over centuries. Climate change poses a serious threat to agriculture, particularly in developing states in Asia and India. Both climate variability and extreme climatic events are affecting agricultural productivity and food security in Asian and Indian states. Moreover, climate change impacts are affecting the livelihoods of common people and aggravating global poverty. Millions of poor across the Asia and India derive their liveli-hoods from common property and natural resource bases like land, water, fisheries, and forestry.

Are bio-fuels a solution to climate change or cause of global hunger? Rsearchers have concluded that shift to bio-fuels is a mistake. It will lead to clearing forests to plant crops for fuel and this will release gases. It is said that increasing production of bio–fuels to combat climate change will release between 2-9 times more carbon gases over the next 30 years than fossil fuels developed states should re-direct bio-fuel subsidies towards preventing deforestation for a better climate outcome.

Clearing forests produces an immediate release of carbon gases into the atmosphere followed by loss of habitats, wildlife and livelihoods. The demand on arable land cannot be met in the EU or the US to plant bio-fuel crops. There is high possibility to shift the burden on land in developing states, *e.g.*, Brazil, Paraguay, Indonesia, which have huge deforestation programmes to supply the world bio-fuel market.

Any delay in addressing climate change through reduction of greenhouse gas emissions will increase the cost of both adaptation and protecting development. Severe food insecurity and poverty in many states is likely to deteriorate significantly as a result of climate change. Climate change, agriculture and food security, natural resources and environmental sustainability, economic growth and poverty reduction are interdependent and this interdependence needs to be at the core of comprehensive policy planning and an integrated resource allocation and implementation strategy, both spatially and temporally.

DEVELOPMENT PARADIGMS AFFECTINGAGRICULTURE AND LIVELIHOODS

Small-holder agriculture in India has consistently under-performed, for reasons that remain only partly understood, despite a succession of theoretical paradigms and analytical frameworks that have been translated into policy prescriptions which have similarly failed to deliver sustained and significant increases in agricultural yields.

For example, the radical reforms implemented under agricultural liberalization programmes throughout India in the 1980s and 1990s were grounded in a theory that agriculture was being stifled by excessive interventionism by the state in agricultural production and marketing, while farmers and traders were undermined by unsustainable input and output subsidies. The shift from "state-led" to "market-led" agriculture achieved disappointing results, leading to a reassessment of both the diagnosis and policy prescriptions. There were several explanations why agricultural liberalization failed in India. These ranged from deficits in the necessary enabling environment, infrastructure like information systems and markets. The states had also withdrawn from research and extension services.

However, in other developing states like in Asia, the states withdrew from agriculture after the necessary investments had been made in infrastructure and market development.

An alternative explanation was that liberalization failed because it was imperfectly and incompletely implemented. Another view was that small-holder agriculture in India has under-performed because of lack of protection or recognition of small-holder farmers by the policy makers and the international bodies like the UN during their "modernization" phase. This protection was necessary to avoid unfair competition from subsidized products from wealthy industrialized states. After a number of debates, certain policies have been put in place to revitalize the agriculture sector in India and especially SSA. They include:

- Improve access to markets and develop modern market chains;
- Assist subsistence farmers to enter markets, and foster sustainable resource management;
- Achieve food security and improved livelihoods for those who remain as subsistence farmers, including improving the resilience of farming systems to climate change; and
- Capitalize on agricultural growth to develop the rural non-farm sector.

Some steps to achieve these outcomes have been suggested as including improvement in agriculture market chains, expanding exports, increasing uptake of yield-stabilizing technologies, and reducing weather risk.

AGRICULTURAL TECHNOLOGY

- To satisfy the increasing needs of food security in India, especially SSA, substantial technological innovation should take place. To increase animal and crop production, modern biotechnology tools of recombinant DNA, including genetic engineering, offer some opportunities for generating such innovation.
- Though the use of some biological technologies has met resistance on social and ethical requirements, biotechnology offers the best solution to reduce hunger in India. Genetically engineered products like GM crops has met with fierce resistance, particularly in Europe on ethical grounds and on concerns of perceived negative impacts of

GM crops on the environment and food safety. Drugs and vaccines have not stirred much controversy.
- Ethical considerations revolve around topics such as the "unnatural" nature of gene transfers across species; possible socio-economic impacts of widening the gap between the rich and poor farmers and states and the fear that agricultural biotechnology will increase the dependency of global food supply on few multinational corporations controlling the seed industry.
- With all the negative publicity of GM crops, there are concerns that the resistance to GM crops by consumers in Europe may have hindered the transfer of this new innovation to the developing states where increasing crop productivity is most urgent.

Despite this resistance, biotechnology is already in play and necessities of life like food, feed, fibre, fuel and medical drugs are being produced. In agriculture, biotechnology tools have been used for animal and plant disease diagnostics, for production of recombinant vaccines against animal diseases and for the improvement of livestock and crops. There has been some lack of support for the development of GMOs in the fight for food security in India from some sectors.

However, National Academies of Sciences called for a concerted effort by all sectors, public and private, to develop GM crops, especially food staples that will benefit consumers and poor farmers in developing nations. It called for the sharing of GM technology developed by private corporations for use in hunger alleviation and to enhance food security in developing states.

The role of biotechnology involves the use of living organisms for human benefit.

It consists of two components:
- Tissue and cell culture and
- DNA technologies including genetic engineering.

Both components are essential for the production of GM plants and animals. Plant tissue and cell culture are relatively low cost technologies which are simple to learn, easy to apply and widely practiced in many developing states. Plant tissue culture aids crop improvement through a range of actions, including:
- Mass propagation of elite stock;
- The provision of virus-free stock through in vitro culture of meristem;
- The selection and generation of somaclonal variants with desirable traits;
- The overcoming of reproductive barriers and the transfer of desirable traits from wild relatives to crops by wide crosses;
- The facilitation of gene transfers using plant protoplast fusions;

- Anther culture to obtain homozygous lines in a breeding programme; and
- *In vitro* conservation of plant germplasm.

The second component of biotechnology, *i.e.* DNA technologies, including genetic engineering, utilizes newly emerging knowledge of the genes and the genetic code to improve crops, trees, livestock and fish. In order to fully realise the benefits of agricultural biotechnology to contribute to poverty alleviation and equitable growth in developing nations, it is crucial that a concerted effort be made to ensure that the benefits of biotechnology are available to a broad spectrum of small farmers in a range of developing states.

For example, the sharing of the rice genome sequence data with researchers in developing states by private sector and the recent announcement of the sequencing of the banana genome are steps in the right direction towards using biotechnology for the benefit of developing nations. Also, there is the question of assisting in capacity building in biosafety in compliance with the recently agreed Cartagena Protocol on biosafety regulations.

This is because the owners of proprietary technologies would be very reluctant to licence their technologies in states that have no biosafety legislation and controls in place.

The development of biosafety in developing states is also important in cases where these states may have to import GMOs for use in breeding programme/ testing in their environment and/or for use as food and feed. Facilitate access to proprietary technologies to developing states. The issue of Intellectual Property Rights (IPR) needs to be addressed. It is important that capacity building in IPR is stimulated in order to meet the minimum requirements of the WTO-TRIPS.

ORGANIC OR INORGANIC MATERIAL OF FERTILIZER

Fertilizer is any organic or inorganic material of natural or synthetic origin that is added to a soil to supply one or more plant nutrients essential to the growth of plants. A recent assessment found that about 40 to 60 per cent of crop yields are attributable to commercial fertilizer use. They are essential for high-yield harvest: European fertilizer market is expected to grow to □15.3 billion by 2018.

Mined inorganic fertilizers have been used for many centuries, whereas chemically synthesized inorganic fertilizers were only widely developed during the Industrial Revolution.

Increased understanding and use of fertilizers were important parts of the pre-industrial British Agricultural Revolution and the industrial Green Revolution of the 20th century. Inorganic fertilizer use has also significantly supported global population growth—it has been estimated that almost half the people on the Earth are currently fed as a result of synthetic nitrogen fertilizer use.

Fertilizers typically provide, in varying proportions:

- *Six Macronutrients:*
  - Nitrogen (N),
  - Phosphorus (P),
  - Potassium (K),
  - Calcium (Ca),
  - Magnesium (Mg), and
  - Sulfur (S).
- *Seven Micronutrients:*
  - Boron (B),
  - Chlorine (Cl),
  - Copper (Cu),
  - Iron (Fe),
  - Manganese (Mn),
  - Molybdenum (Mo), and
  - Zinc (Zn).

The macronutrients are consumed in larger quantities and are present in plant tissue in quantities from 0.15 per cent to 6.0 per cent on a dry matter (0 per cent moisture) basis (DM). Micronutrients are consumed in smaller quantities and are present in plant tissue on the order of parts per million (ppm), ranging from 0.15 to 400 ppm DM, or less than 0.04 per cent DM. Only three other macronutrients are required by all plants: carbon, hydrogen, and oxygen. These nutrients are supplied by water and carbon dioxide. The nitrogen-rich fertilizer ammonium nitrate is also used as an oxidizing agent in improvised explosive devices, sometimes called *fertilizer bombs*, leading to sale regulations.

LABELLING AND FORMS

The labelling of fertilizers varies. In most countries the macronutrients are labeled with an *NPK* analysis. The three numbers on the fertilizer label represent an analysis of the composition by weight. These three numbers correspond to nitrogen, phosphorus, and potassium and always appear in that specific order. When a 4th number is included, it indicates the sulfur content.

While the number for "N" represents the percentage weight of nitrogen, in some European countries, the other two components are not for the analysis of the element, but rather, the analysis of the "available" or "soluble" form of the element. In traditional chemical analysis, the tests used treated the sample so as to measure the equivalent P_2O_5 and K_2O. For instance, some potassium-bearing rocks do not count as having available potassium. Similarly the number for "P" in some countries is actually the weight of an equivalent quantity of P_2O_5 whereas in others it refers to elemental phosphorus. In order to calculate the weight of P in the formulation, the weight of P_2O_5 can be multiplied by 0.44 to compensate for the weight of the oxygen in the molecule. For example, a bag of 10-10-10 has 10 pounds of nitrogen, 10 pounds of P_2O_5, but only 4.4 pounds of P. Likewise, the number for "K" in Europe can refer to the weight of an equivalent quantity of K_2O, whereas in Australia it refers to elemental potassium.

In order to calculate the weight of K in the formulation, the weight of K_2O can be multiplied by 0.83 to compensate for the weight of the oxygen in the molecule. For example, a bag of 10-10-10 has 10 pounds of K_2O, but only 8.3 pounds of K. As an example, the fertilizer *potash* is composed of 52 per cent potassium and 48 per cent chlorine by weight; chemical analysis of 100g of potassium chloride, would show 63g of equivalent potassium oxide when done in the manner of fertilizer analysis. The *percentage yield* of K_2O from the original 100g of fertilizer is the number shown on the label. A potash fertilizer would thus be labelled 0-0-63, and *0-0-52.*

Forms

Fertilizers come in various forms. The most typical form is solid fertilizer in granulated or powdered. The next most common form is liquid fertilizer; some advantages of liquid fertilizer are its immediate effect and wide coverage. There are also slow-release fertilizers which reduce the problem of "burning" the plants due to excess nitrogen. Polymer coating of fertilizer ingredients gives tablets and spikes a "true time-release" or *"staged nutrient release"* (SNR) of fertilizer nutrients. More recently, organic fertilizer is on the rise as people are resorting to environmental friendly products. Although organic fertilizer usually contain less nutrients, some people still prefer organic due to natural ingredients.

INORGANIC COMMERCIAL FERTILIZER

Fertilizers are broadly divided into organic fertilizers or inorganic or commercial fertilizers. Plants can only absorb their required nutrients if they are present in easily dissolved chemical compounds. Both organic and inorganic fertilizers provide the same needed chemical compounds. Organic fertilizers provided other macro and micro plant nutrients and are released as the organic matter decays—this may take months or years. Organic fertilizers nearly always have much lower concentrations of plant nutrients and have the usual problems of economical collection, treatment, transportation and distribution.

Inorganic fertilizers nearly always are readily dissolved and unless added have few other macro and micro plant nutrients. Nearly all nitrogen that plants use are in the form of NH_3 or NO_3 compounds. The usable phosphorus compounds are usually in the form of phosphoric acid (H_3PO_4) and the potassium (K) is typically in the form of potassium chloride (KCl). In organic fertilizers nitrogen, phosphorus and potassium compounds are released from the complex organic compounds as the animal or plant decays.

In commercial fertilizers the same required compounds are available in easily dissolved compounds that require no decay—they can be used almost immediately after water is applied. Inorganic fertilizers are usually much more concentrated with up to 64 per cent of their weight being a given plant nutrient, compared to organic fertilizers that typically often have only 0.4 per cent or less of a their weight a given plant nutrient.

Nitrogen fertilizers are often made using the Haber-Bosch process which uses natural gas ($CH_{4+)}$ for the hydrogen and nitrogen gas (N_2) from the air at an elevated temperature and pressure in the presence of a catalyst to form ammonia (NH_3) as the end–product. This ammonia is used as a feedstock for other nitrogen fertilizers, such as anhydrous ammonium nitrate (NH_4NO_3) and urea ($CO(NH_2)_2$). These concentrated products may be diluted with water to form a concentrated liquid fertilizer. Deposits of sodium nitrate ($NaNO_3$) are also found the Atacama desert in Chile and was one of the original nitrogen rich inorganic fertilizers used.

It is still mined for fertilizer. In the Nitrophosphate process or Odda Process, phosphate rock with up to a 20 per cent phosphorus (P) content is dissolved with nitric acid (HNO_3) to produce a mixture of phosphoric acid (H_3PO_4) and calcium nitrate ($Ca(NO_3)_2$). This can be combined with a potassium fertilizer to produce a *compound fertilizer* with all three N:P:K: plant nutrients in easily dissolved form.

Phosphate rock can also be processed into water-soluble phosphate (P_2O_5) with the addition of sulfuric acid (H_2SO_4) to make the phosphoric acid in phosphate fertilizers. Phosphate can also be reduced in an electric furnace to make high purity phosphorous; however, this is more expensive than the acid process. Potash can be used to make potassium (K) fertilizers. All commercial potash deposits come originally from marine deposits and are often buried deep in the earth.

Potash ores are typically rich in potassium chloride (KCl) and sodium chloride (NaCl) and are obtained by conventional shaft mining with the extracted ore ground into a powder. For deep potash deposits hot water is injected into the potash which is dissolved and then pumped to the surface where it is concentrated by solar induced evaporation. Amine reagents are then added to either the mined or evaporated solutions. The amine coats the KCl but not NaCl.

Air bubbles cling to the amine + KCl and float it to the surface while the NaCl and clay sink to the bottom. The surface is skimmed for the amine + KCl which is then dried and packaged for use as a K rich fertilizer—KCl dissolves readily in water and is available quickly for plant nutrition. Compound fertilizers often combine N, P and K fertilizers into easily dissolved pellets. The N:P:K ratios quoted on fertilizers give the weight per cent of the fertilizer in nitrogen (N), phosphate (P_2O_5) and potash (K_2O equivalent). The use of commercial inorganic fertilizers has increased steadily in the last 50 years, rising almost 20-fold to the current rate of 100 million tonnes of nitrogen per year. Without commercial fertilizers it is estimated that about one-third of the food produced now could not be produced. The use of phosphate fertilizers has also increased from 9 million tones per year in 1960 to 40 million tones per year in 2000. A maize crop yielding 6–9 tones of grain per hectare requires 31–50 kg of phosphate fertilizer to be applied, soybean requires 20–25 kg per hectare. Yara International is the world's largest producer of nitrogen based fertilizers.

Controlled-release Types

Urea and formaldehyde, reacted together to produce sparingly soluble polymers of various molecular weights, is one of the oldest controlled-nitrogen-release technologies, having been first produced in 1936 and commercialized in 1955. The early product had 60 per cent of the total nitrogen cold-water-insoluble, and the unreacted less than 15 per cent. Methylene ureas were commercialized in the 1960s and 1970s, having 25 and 60 per cent of the nitrogen cold-water-insoluble, and unreacted urea nitrogen in the range of 15 to 30 per cent.

Isobutylidene diurea, unlike the methylurea polymers, is a single crystalline solid of relatively uniform properties, with about 90 per cent of the nitrogen water-insoluble. In the 1960s the National Fertilizer Development Centre began developing Sulfur-coated urea; sulfur was used as the principle coating material because of its low cost and its value as a secondary nutrient.

Usually there is another wax or polymer which seals the sulfur; the slow release properties depend on the degradation of the secondary sealant by soil microbes as well as mechanical imperfections in the sulfur. They typically provide 6 to 16 weeks of delayed release in turf applications. When a hard polymer is used as the secondary coating, the properties are a cross between diffusion-controlled particles and traditional sulfur-coated. Other coated products use thermoplastics to produce diffusion-controlled release of urea or soluble inorganic fertilizers.

"Reactive Layer Coating" can produce thinner, hence cheaper, membrane coatings by applying reactive monomers simultaneously to the soluble particles.

"Multicote" is a process applying layers of low-cost fatty acid salts with a paraffin topcoat. Besides being more efficient in the utilization of the applied nutrients, slow-release technologies also reduce the impact on the environment and the contamination of the subsurface water.

Application

Synthetic fertilizers are commonly used for growing all crops, with application rates depending on the soil fertility, usually as measured by a soil test and according to the particular crop. Legumes, for example, fix nitrogen from the atmosphere and generally do not require nitrogen fertilizer. Studies have shown that application of nitrogen fertilizer on off-season cover crops can increase the biomass of these crops, while having a beneficial effect on soil nitrogen levels for the main crop planted during the summer season.

Nutrients in soil can be thrown out of balance with high concentrations of fertilizers. The interconnectedness and complexity of this soil "food web" means any appraisal of soil function must necessarily take into account interactions with the living communities that exist within the soil.

Stability of the system is reduced by the use of nitrogen-containing fertilizers, which causesoil acidification. Applying excessive amounts of fertilizer has negative environmental effects, and wastes the growers' time and money.

To avoid over-application, the nutrient status of crops should be assessed. Nutrient deficiency can be detected by visually assessing the physical symptoms of the crop. Nitrogen deficiency, for example has a distinctive presentation in some species. However, quantitative tests are more reliable for detecting nutrient deficiency before it has significantly affected the crop. Both soil tests and plant tissue tests are used in agriculture to fine-tune nutrient management to the crops needs.

Problems with Inorganic Fertilizer

Common agricultural grade phosphate fertilizers usually contain impurities such as fluorides, cadmium and uranium, although concentrations of the latter two heavy metals are dependent on the source of the phosphate and the production process. These potentially harmful impurities can be removed; however, this significantly increases cost.

Highly pure fertilizers are widely available and perhaps best known as the highly water soluble fertilizers containing blue dyes used around households. These highly water soluble fertilizers are used in the plant nursery business and are available in larger packages at significantly less cost than retail quantities. There are also some inexpensive retail granular garden fertilizers made with high purity ingredients.

Oregon and Washing in US have fertilizer registration programmes with on-line databases listing chemical analyses of fertilizers. The most widely used inorganic fertilizer is super-phosphate and its double and triple strengthed derivatives double super and triple super. Superphosphate was first developed by Lawes at the Rothamstead Agricultural Research Institute in England in the early 19th century. Lawes added sulfuric acid to conventional rock phosphate containing the mineral apatite, a calcium fluoro-phosphate.

The resulting water soluble phosphorus was able to significantly improve yields on a variety of crops at the Rothamstead Centre and the Superphosphate industry was born. Unfortunately over decades of subsequent useage-it became clear that the solubilisation of fluorine also occurred in the process and this had the same effect as the other halogen sterilants over time-soil sterilization.

Effectively farmers unknowingly became 100 per cent dependent on "bought in" water soluble, inorganic fertilizers since the sterilization of soil microflora including its micorhizza, reduced the availability of other natural and trace minerals within the soil. This to some extent explains the resurgence of interest in organic and particularly "biodynamic" farming systems since these systems replace the essential soil organisms so essential to converting soil minerals into plant available nutrients. They do this by a variety of processes including chelation whereby essential minerals become plant available–as measured by weak citric acid extraction techniques.

Hence the citric acid solubility of phosphate rocks has emerged as a measure of plant availability and enabled so-called "reactive" phosphate rocks to be used as fertilizer minerals. These should not be confused with high fluorine apatite rocks in which the fluoride content performs a similar function to its role in hardening teeth enamel, *i.e.*, immobilizing phosphorus. This explains the oceanic origins of many of these high fluorine rocks since the fluorine absorbed from the sea has prevented what were originally massive deposits of bird guano-from being leached from the coral based limestone rocks on which they were originally deposited.

Also regular use of acidulated fertilizers generally contribute to the accumulation of soil acidity in soils which progressively increases aluminium availability and hence toxicity. The use of such acidulated fertilizers in the tropical and semi-tropical regions of Indonesia and Malaysia has contributed to soil degredation on a large scale from aluminium toxicity, which can only be countered by applications of limestone or preferably magnesian dolomite, which neutralises acid soil pH and also provides essential magnesium.

Trace Mineral Depletion

Many inorganic fertilizers, particularly those based on superphosphate, may not replace trace mineral elements in the soil which become gradually depleted by crops. This depletion has been linked to studies which have shown a marked fall in the quantities of such minerals present in fruit and vegetables.

Explanations for this include the early encouragement of so-called "luxury consumption" of trace elements as a result of their acidulation and subsequent dissolution in soil water, by free sulphuric acid sourced from superphosphate. This mechanism has also been identified as a possible causal agent for take-up of the heavy metal cadmium from superphosphate based fertilizers. In Western Australia deficiencies of zinc, copper, manganese, iron and molybdenum were identified as limiting the growth of broad-acre crops and pastures in the 1940s and 1950s. Such nutrients are described as "rate limiting" nutrients.

Soils in Western Australia are very old, highly weathered and deficient in many of the major nutrients and trace elements. Since this time these trace elements are routinely added to inorganic fertilizers used in agriculture in this state. Many soils around the world are deficient in zinc, leading to deficiency in plants and humans.

Over-fertilization

Over-fertilization of a vital nutrient can be as detrimental as under–fertilization. "Fertilizer burn" can occur when too much fertilizer is applied, resulting in a drying out of the leaves and damage or even death of the plant. Fertilizers vary in their tendency to burn roughly in accordance with their salt index.

High Energy Consumption

In the USA in 2004, 317 billion cubic feet of natural gas was consumed in the industrial production of ammonia, less than 1.5 per cent of total US annual consumption of natural gas. A 2002 report suggested that the production of ammonia consumes about 5 per cent of global natural gas consumption, which is somewhat under 2 per cent of world energy production. Ammonia is overwhelmingly produced from natural gas, but other energy sources, together with a hydrogen source, can be used for the production of nitrogen compounds suitable for fertilizers.

The cost of natural gas makes up about 90 per cent of the cost of producing ammonia. The increase in price of natural gases over the past decade, along with other factors such as increasing demand, have contributed to an increase in fertilizer price.

Long-term Sustainability

Inorganic fertilizers are now produced in ways which theoretically cannot be continued indefinitely by definition as the resources used in their production are non-renewable. Potassium and phosphorus come from mines and such resources are limited. However, more effective fertilizer utilization practices may decrease present usage from mines.

Improved knowledge of crop production practices can potentially decrease fertilizer usage of P and K without reducing the critical need to improve and increase crop yields. Atmospheric nitrogen is effectively unlimited but this is not in a form useful to plants. To make nitrogen accessible to plants requires nitrogen fixation.

Artificial nitrogen fertilizers are typically synthesized using fossil fuels such as natural gas and coal, which are limited resources. In lieu of converting natural gas to syngas for use in the Haber process, it is also possible to convert renewable biomass to syngas to supply the necessary energy for the process, though the amount of land and resources necessary for such a project may be prohibitive.

ROLE OF INFORMATION TECHNOLOGY IN AGRICULTURE

The agriculture scenario all over the world is undergoing a rapid change particularly after WTO Agreements came to existence. In order to take full advantage of the changing global agricultural scenario interconnecting of policies related to pricing, marketing and trading of agricultural commodities are also reviewed. Simultaneously there is also need to review and revitalise the mechanism for transfer of technology under changing environment. It is readily accepted that increased information flow has a positive effect on the agricultural sector and individual firms.

However, collecting and disseminating information is often difficult and costly. Information Technology (IT) offers the ability to increase the amount of information provided to all participants in the agricultural sector and to decrease the cost of disseminating the information. An understanding of the factors associated with IT adoption and use in agriculture will enable the development of strategies to promote IT adoption and increase the effectiveness and efficiency of information used in agriculture. It is a fact that access to information holds the key for successful development.

Improved communications and information access is directly related to socio-economic development of any nation. Agriculture is one of the prospective areas in which IT can effectively be applied particularly for the social and economic development of the Indian agrarian community. However, rural population in our country still have difficulties in accessing crucial information in the forms they can understand in order to make timely decisions for better farming.

IT is generating possibilities to solve such problems of different categories of end users. For this purpose electronic communications infrastructure needs to be established in the country for remote rural areas. The challenge is not only to improve the accessibility of communications technology to the rural population but also to improve the relevance of information to local development. The present object depicts the changing scenario of information dissemination exploiting the information technology to the farmers for their agricultural development.

IT FOR AGRICULTURAL PRODUCTION AND MARKETING

IT is playing an important and vital role in agricultural production and marketing. IT allows farmers to save time on order and delivery and getting feedback. In the existing competition, there is a need to rapidly attract new customers as well as retain existing customers.

In order to take the real status of agricultural production and marketing, there is an urgent need to develop the following items:

- *Farmers' Crop Database must be Managed*: The database includes the kinds of crops, the size of cultivated area, time of harvest and yield. Farmers or the extension personnel transmit those data via the Internet to database server. Further, information provides the farmer with an important instrument for decision making and taking action.
- *Crops Information Service System should be Created:* This system analyses the crop data to create some statistical tables. Farmers can access these statistical data by browsing the homepage and make their production plan. Changes within the structure of agriculture will probably have an impact on the selection and types of acquisition of software and other integrated systems made by the farmers.
- *Production Techniques and Information Enquiry System should be Created*: This system integrates the production techniques and information, which are developed by experimental agricultural institutes and agricultural improvement stations. Farmers can find out relevant production information through this enquiry service system.
- *Production Equipment's Enquiry Service System should be Created*: This system gathers information from the companies of seeds and crop production equipment to build the production equipment's enquiry service system. At the same time, allow relevant companies to access this system and enter their own data. Therefore, farmers can order the needed items through this system.

Information is critical to the social and economic activities that comprise the development process. Development economy has witnessed for revolutions in agricultural, bio-technological, industrial and information technology. Good communication system and information system reinforce commitments to sustainable productivity. The Government of India is giving more thrust on agriculture, food and information technology sectors towards achievement of economic reforms to achieve high growth rate in production in the years to come. The National Agriculture Policy announced addresses the challenges arising out of economic liberalization and globalization.

It seeks to actualize the vast untapped growth potential of Indian agriculture, strengthen rural infrastructure to support faster agricultural development, promote value addition and secure a fair standard of living for farmers and farm workers. The National Agriculture Policy lays emphasis on the use of Information Technology for achieving a more rapid development of agriculture in India.

In pursuance thereof, the Department of Agriculture and Cooperation (DAC) has formulated information technology (IT) Vision 2020. This vision *inter alia* envisages that:

- Information relating to agriculture sector would be available to the ultimate users—the farmers—for optimizing their productivity and income;
- Extension and advisory services making use of information technology would be available to the farmers on round the clock basis;
- The tools for information technology will provide networking of agriculture sector not only in the country but also globally and the Union and State Government Departments will have reservoirs of data base; and
- The long term vision on "Information Technology in Agriculture Sector" is to bring farmers, researchers, scientists and administrators together by establishing "Agriculture on-line" through exchange of ideas/information.

In future, information technology will reduce the cost and time of information system. IT will bring new information services for agricultural development that will enable the farmers to have much greater control over the information channels.

AGRICULTURAL IMPLICATIONS OF IT

Information technology provides answers to a number of questions to the farmers. For example, what are the benefits of more irrigation? Is it cost-effective to apply additional chemicals? When is the best time to sell crops or buy inputs? With improved record-keeping, more detailed cost analysis and more sophisticated marketing strategies, farmers are making better decisions and earning higher profits.

The Internet is increasing communication and business opportunities within the agricultural community, which previously operated in the relative isolation of rural areas. Farmers, agricultural researchers, cooperatives, suppliers and buyers use the Internet to exchange ideas and information, as well as to conduct business with each other. Machinery, seed chemicals and other types of agricultural products can be purchased and sold online. People can search for jobs and employees. It is to be noted that, the farmer is in no position to use IT directly.

The literacy levels, language barrier as most of the application software are predominantly in English, cost of computers, poor communication infrastructure make it impossible for individual farmers, particularly small farmers to directly adopt IT. This calls for institutional effort to harness to create the necessary IT based services to farmers. But in India, one prominent problem is that most of the farmers own small holdings, this seems to be difficult. In this situation, it may be made possible by adopting the corporate farming system, which is the need of the hour with advent of new agricultural policy.

By taking up corporate farming, a group of farmers can install a computer and any educated young man from that group can undergo training of how to browse the Internet. He can provide the farmers current commodity, analysis reports on world markets and trade for different commodities. Food market overviews provide valuable information about some of the most important export market. IT can help to provide the information on the likely price distribution of key commodities over the coming years. Such information helps farmers and traders make decisions on when and in what ways to market their grain. Whether, to sell at harvest or store on-farm in anticipation of higher later in the season. When combined with enterprise budget data, the information can also be used in deciding which crops to produce in the coming season.

In order to encourage farmers to obtain best possible price, information on various agricultural output markets is also being provided. The objective of this activity is to provide status of price at different markets to facilitate farmer to move his produce to the market where he can expect better price. The entire exercise will not be useful unless necessary arrangements are made to ensure that the farmers utilize this facility. The contribution of information technology in bringing down costs, increasing efficiency and improving productivity and thereby contributing to the bottom line needs no special emphasis.

In the fertilizer marketing context, IT can play a major role in efficient sales operations, checking the marketing costs, safeguarding market share and providing efficient customer services. A well conceived IT set up can endow decision makers at all levels with better reflexes to effectively respond to market conditions. IT helps producers monitor and respond to weather variability on a day-to-day basis.

Solar-powered weather stations in the field can be hooked up to a farmer's computer to relay information about current air and soil temperature, precipitation, relative humidity, leaf wetness, soil moisture, day length, wind speed and solar radiation. Producers use the Internet to monitor prices quickly and as often as they like. Farmers from around the world can exchange ideas, post questions and get answers about specific topics. Thus, it is said that the importance of information technology in the field of agriculture is emerging.

The challenges of cost intensive, highly technical agricultural technologies are knocking the door to the farmers which is ignited by the globalization. There is significant shift from agriculture supply driven to demand driven paradigm in new emerging and changing economics policy. It is viewed that future agricultural growth would be information driven. New information must reach to the ultimate user at the fastest speed to harness its potential benefits. Information like seed, water, nutrients and plants protection biological is one of the key inputs for successful farming. Knowledge intensive and precision farming techniques will be the guiding lines for sustainable agriculture in the future.

10

Characteristics, Costs, and Issues

Organic milk production has been one of the fastest growing segments of organic agriculture in the United States. Between 2000 and 2005, the number of certified organic milk cows on U.S. farms increased by an annual average of about 25 per cent, from 38,000 to more than 86,000. Many of these cows were on small dairy operations that had switched to organic production to improve farm profitability. While organic milk cows accounted for about 1 per cent of all U.S. milk cows in 2005, the sector continues to grow.

USDA estimates indicate that organic milk increased from 2 per cent of U.S. fluid milk product sales in 2006 to 3 per cent in 2008. Certified organic milk production systems rely on ecologically based standards that prohibit the use of antibiotics and hormones in the cow herd and the use of synthetic chemicals in dairy feed production. Certified organic milk production systems also attempt to accommodate the animals' natural nutritional and behavioural requirements, such as ensuring that cows have access to pasture.

These requirements add to production costs and create challenges for widespread adoption, such as higher managerial costs and risks of shifting to a new way of farming, and significant time and costs associated with the transition to organic production. This report describes organic dairy farms in the United States, including information about farm size, location, production costs, and the types of technologies used, and discusses the challenges of organic milk production. Although the analysis focuses on organic dairy farms, a comparison with conventional dairy farms adds value to the discussion.

Organic milk producers usually begin as conventional dairy operators who then go through what can be a challenging and costly transition process. During the transition, organic dairies must change their animal husbandry, land, and crop management, source new and different inputs, and initiate the certification process. To become a certified organic dairy under current standards of USDA's National Organic Programme, the pasture and cropland providing feed for organic dairies must be managed organically for a minimum of 36 months.

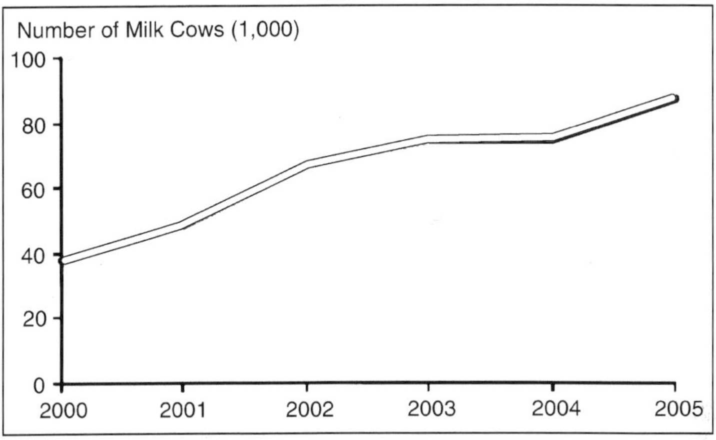

Fig. Number of Organic Milk Cows on U.S. Farms

Notes:

Between 2000 and 2005, the number of organic milk cows increased by an annual average of about 25 per cent

Current standards also require the dairy herd to be fed 100 per cent organic feed and to receive organic health care for 12 months before being certified. Organic dairy animals may not be given hormones, such as rBST, or antibiotics for any reason. Grazing is required for all animals over 6 months of age. These requirements mean that products and feeds that meet organic standards must be found, and organic feeds can be priced significantly higher than conventional feeds.

Also, the approach to management must be adjusted on most farms as many conventional inputs, such as antibiotics, are no longer available. The certification process can also be time consuming and tedious as farmers must develop an organic systems plan that describes practices and substances used in production. In addition, detailed production records must be kept for 5 years post-certification for a farm to be in compliance with the regulations, and access to these records must be provided to USDA and its certifying agents.

USDA's *National Organic Programme* (NOP) rules state that access to pasture must be provided for ruminant animals, but do not indicate how much pasture should be allocated or how much dairy feed should be provided by pasture.

The most common technology used on conventional dairy operations confines milk cows in large barns and limits access to pasture. Therefore, interpretation of the pasture rule, largely left to individual certifiers, has important implications for land requirements and costs of organic milk production for farmers considering the transition.

Partly in response to comments, complaints, and non-compliance regarding pasture use on organic dairy operations, USDA proposed a rule in October 2008 to clarify and strengthen the NOP pasture requirement. This proposed rule would amend livestock and related provisions of the NOP, providing better consumer assurances that the USDA organic label on dairy products means that cattle graze on pasture during the pasturegrowing season. Milk prices paid to U.S. producers are determined by a complex system.

A minimum, or base price, is determined by a combination of the dairy price support programme and Federal milk marketing orders. Quality adjustments are made to the minimum price based on the content of fats and other milk solids and the volume of milk sold, among other factors. Organic milk pricing is subject to the same minimum as all other milk, but processors pay a premium for the value added by the organic brand. Also, organic producers with cows using pasture as their primary forage may receive higher prices because the milk tends to have higher fat and other solids content.

Organic milk producers are paid a fixed price per hundredweight (cwt) for organic milk, according to annual contracts with organic creameries that purchase the milk. In contrast, conventional milk producers are paid a blend price determined by spot markets for butter, cheese, non-fat dry milk, and whey, which may vary dramatically each month. The relative price stability of organic milk, compared with conventional milk, is a benefit for producers who transition to organic.

DAIRY FARM DATA

Dairy farm data used in this study came from USDA's 2005 Agricultural Resource Management Survey (ARMS) of U.S. milk producers. The ARMS, conducted annually by the National Agricultural Statistics Service (NASS) and Economic Research Service (ERS), provides detailed farm information on income, expenses, assets, and debt, as well as farm and operator characteristics.

The 2005 ARMS included a version that highlights milk production practices and costs. This version targets dairy operations in 24 States that accounted for more than 90 per cent of national milk production and covered all major production areas. The surveyed dairy farms were chosen from a list of farm operations maintained by USDA's National Agricultural Statistics Service (NASS). The target population of the dairy version included farms that were in the dairy business for all 12 months of 2005 and milked at least 10 cows at any time during the year.

The dairy survey collected information to estimate average milk production costs. Screening out farms that milked fewer than 10 cows excluded farms with dairy cows for onfarm consumption and other non-commercial activities, such as youth projects.

A subsample that targeted organic operations was also part of the 2005 ARMS dairy version. Of the total 2,987 dairy farms sampled, 737 samples were targeted at organic operations in 19 States as identified by the major organic milk processors and certifiers.

After accounting for non-response and missing data, information on 1,787 farms, including 352 certified organic milk operations in 14 States, were available from the ARMS. Among the organic dairies, 325 sold more than 90 per cent of milk production as certified organic, 18 were transitioning to organic production, and 9 were mixed operations. Each surveyed farm represented a number of similar farms in the population as indicated by the surveyed farm's expansion factor.

The expansion factor, or survey weight, was determined from the farm's selection probability, expanding the sample to represent the target population. Farm survey weights ensured that the sample expanded to represent dairy operations in the 24 States and that organic operations represented an appropriate share of the population despite a disproportionate representation in the sample.

Organic farms accounted for 18 per cent of the sample, but only 2 per cent of the weighted number of farms.

ECONOMIES OF SIZE IN ORGANIC MILK PRODUCTION

The emergence of large conventional dairy farms and the continued shift of production towards such farms suggest that economies of size play a major role in the structure of dairy farming. Economies of size and the role it takes in the structural changes observed in conventional milk production were documented by MacDonald *et al.* This section reports on the relationship between operation size and production costs in organic milk production, including what size and costs might mean for the structure of the industry.

To evaluate the relationship between organic milk production costs and operation size, surveyed producers were divided into groups by size, and differences in farm characteristics and practices, production costs and net returns, and labour use were compared among the groups.

The size groups were defined by the largest number of cows milked on the operation during 2005:

- Fewer than 50 cows;
- 50-99 cows;
- 100-199 cows; and
- 200 cows or more.

CHARACTERISTICS AND PRACTICES BY SIZE

Data for 2005 indicated that about 45 per cent of organic dairies milked fewer than 50 cows, while 87 per cent had fewer than 100 cows. These small operations accounted for a disproportionately small share of organic milk cows and production. In contrast, about 13 per cent of organic dairies had 100 cows or more, but they accounted for 44 per cent of milk cows and nearly half of production.

The largest organic dairies—those with 200 cows or more—had an average herd size of almost 500 cows per farm and included nearly a third of organic milk cows. These large dairies had a higher average milk yield per cow than smaller dairies and produced 37 per cent of organic milk.

The most striking difference among organic dairies in each size group was their location. Northeast and Upper Midwest farms accounted for 96 per cent of dairies with fewer than 50 cows. Farms in the West accounted for 80 per cent of dairies with 200 cows or more. Within each region, 97 per cent of organic dairies in the Northeast, 88 per cent of organic dairies in the Upper Midwest, and 85 per cent of organic dairies in the Corn Belt had fewer than 100 cows. In contrast, 83 per cent of organic dairies in the West had 100 cows or more and 63 per cent had 200 cows or more.

The use of several milk production technologies and practices tended to increase with size of organic operation, but most of the differences were between the smallest and other organic farms. For example, use of a milking parlor increased from 22 per cent for the smallest dairies to nearly all dairies with 100 cows or more.

Thirty-eight per cent of the smallest farms and over 50 per cent of other farms participated in the Dairy Herd Improvement programme.

Table. Dairy Farm Characteristics and Production Practices, by Size of Organic Operation, 2005

	Size of Organic Dairy			
Item	Fewer than 50 Cows	50-99 Cows	100-199 Cows	200 Cows or More
Per cent of farms	45	42	8	5
Per cent of milk cows	21	35	13	31
Per cent of milk production	18	33	12	37
Milk cows (number per farm)	37	68	132	490
Milk production (pounds per cow)	11,884	12,796	12,008	16,133
Per cent of farms				
Region:				
Northeast (ME, NY, PA, VT)	51	46	17	0
Upper Midwest (MI, MN, WI)	45	39	52	20

Corn Belt (IL, IN, IA, MO, OH)	4	12	15	0
West (CA, ID, OR, WA)	0	3	16	80
Production practices:				
DHI programme participation[1]	38	51	53	60
Pasture-based feeding[2]	70	60	49	54
Artificial insemination	70	76	71	87
Embryo transplants or sexed semen	3	3	6	2
Controlled breeding/calving season	38	35	36	20
Regular veterinary services	28	41	72	55
Nutritionist services	30	53	66	70
Computerized milking system	1	4	4	7
Computerized feeding system	id	3	5	31
Milking parlor	22	45	88	100
Kept individual cow records	59	64	62	70
Johne's disease programme participation	18	28	44	45
Onfarm computer records	13	25	34	41
Dairy information from Internet	30	46	56	64
Forward-purchased inputs	4	10	22	16
Negotiate input price discounts	18	21	33	36

Notes:

id=Insufficient data for disclosure.

[1]Dairy Herd Improvement.

[2]Pasture-based feeding is defined as providing at least half of the forage fed to milk cows during the grazing months from pasture. Organic dairies reported an average grazing period for milk cows of 6. 5 months in 2005.

Use of regular veterinary and nutritionist services was also much lower on the smallest farms. Likewise, the use of business management tools, such as computer records, the Internet, and forward-purchasing inputs, was less. Pasture-based feeding was used by 70 per cent of the smallest dairies and was also used by more than half of the largest dairies.

PRODUCTION COSTS AND NET RETURNS BY SIZE

Average feed costs differed little among the dairies in each size group, but average operating costs for organic dairies increased from the smallest to largest size group due to the additional costs of hired labour on the largest farms. Total operating costs were about $2 per cwt higher on the largest organic dairies than on the smallest. Once capital costs were added, cost differences among the size groups were negligible, as lower capital costs on the larger operations offset the additional hired labour costs.

The smallest farms, however, used significantly more unpaid labour, mainly from farm operators and other family members, than did larger farms. After adding an opportunity cost for this unpaid labour, total economic costs for organic dairies declined with size and were nearly $14 per cwt less on the largest farms than on the smallest.

Table. Production Costs and Returns, by Size of Organic Dairy Operation, 2005

	Size of Organic Dairy			
Item	**Fewer than 50 Cows**	**50-99 Cows**	**100-199 Cows**	**200 Cows or More**
Dollars per cwt sold				
Feed:	12.96	12.67	11.93	13.06
Purchased feed	5.16	4.94	5.29	8.99
Homegrown harvested feed	6.96	7.00	6.06	3.51
Grazed feed	0.83	0.73	0.58	0.56
Hired labour	0.79	1.75	2.25	3.21
Capital costs	6.02	6.20	5.04	3.58
Opportunity cost of unpaid labour	13.42	7.45	4.84	1.15
Cost summary:				
Operating costs	17.65	18.25	18.10	19.66
Operating and capital costs	23.67	24.44	23.13	23.25
Total economic costs	38.50	33.36	29.13	24.84
Gross value of production	25.57	24.88	23.46	23.59
Returns above total economic costs	-12.93	-8.48	-5.67	-1.26

Notes:

Costs are defined in Appendix A: Measuring Milk Production Costs. The gross value of production includes milk sales, cull and breeding animal sales, revenue from leasing dairy animals or space to other operations, co-op patronage dividends, dairy assessment rebates or refunds, and the value of manure. The value of milk sales was determined from prices reported by the survey respondents.

The relationship between average production costs and operation size for organic dairies was similar to that for conventional dairies among the size groups that characterize the organic industry.

Average production costs for conventional milk production declined from about $30 per cwt on operations with fewer than 50 cows to about $18 per cwt on operations with 200-499 cows. Average costs dropped to less than $14 per cwt on conventional operations with more than 1,000 cows. Lower costs among the largest conventional producers suggest that similar economies of size may be available to organic producers if they increase the size of their operation above the industry norm.

However, the additional costs of complying with pasture requirements and securing organic inputs in large volume may limit the cost advantages for larger organic operations. Net returns to the organic dairy enterprise were computed as the difference between the gross value of production and total economic costs. Average net returns in 2005 were negative for all organic dairy size groups (from-$12.93 per cwt among dairies with fewer than 50 cows to-$1.26 per cwt among dairies with 200 cows or more).

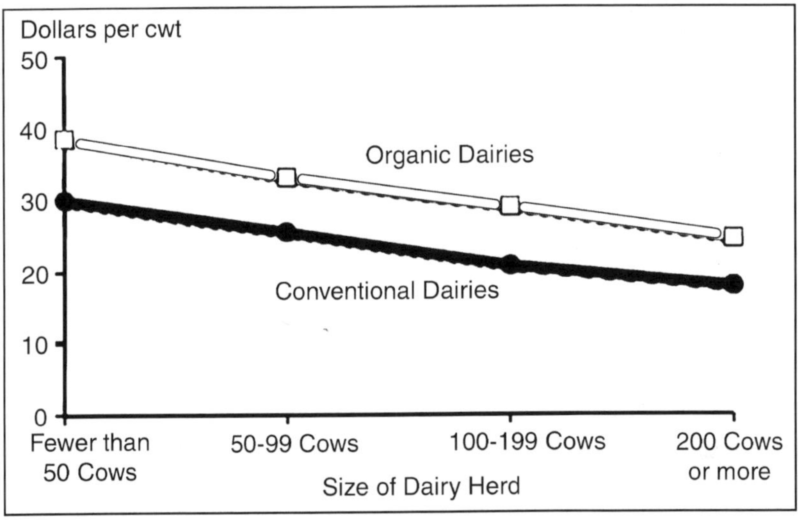

Fig. Total Economic Costs for each Type of Dairy Producers, by Farm Size

Notes:

The relationship between production costs and operation size for organic dairies was similar to conventional diaries, only at higher costs

Costs shown for conventional dairies of 200 cows or more are reported for 200-499 cows by MacDonald *et al.*, with an average size of 295 cows. The average size of organic dairies in the 200 cows or more group is 490 cows.

These averages, however, mask the variation in net returns among dairies. Some operations may be more productive because they are well managed, while others may pay less than average for inputs or receive above-average milk prices, and thus these farms may be profitable while their group, on average, is not. For example, 10 per cent of organic dairies with fewer than 50 cows and almost half of those with 200 cows or more generated positive returns above total economic costs.

Farmers are unlikely to start or expand a dairy operation if they are unable to recover their investment in capital and labour. Farmers who have already made substantial capital investments in dairy facilities and equipment, however, with little or no opportunity for use outside the dairy operation. These costs, therefore, have no bearing on the decision to operate in the short term, only becoming important when these capital assets must be replaced.

To better measure costs relevant to shortterm operating decisions, the share of farms with gross returns exceeding all costs except for capital costs was computed. About a fourth of organic dairies with fewer than 50 cows covered these costs, compared with almost threefourths of those with 200 cows or more.

Some small organic dairy farms appeared to be earning enough to keep operating, but on average, farms in the smaller size groups were not covering the opportunity costs of their investments in capital and the operator's time.

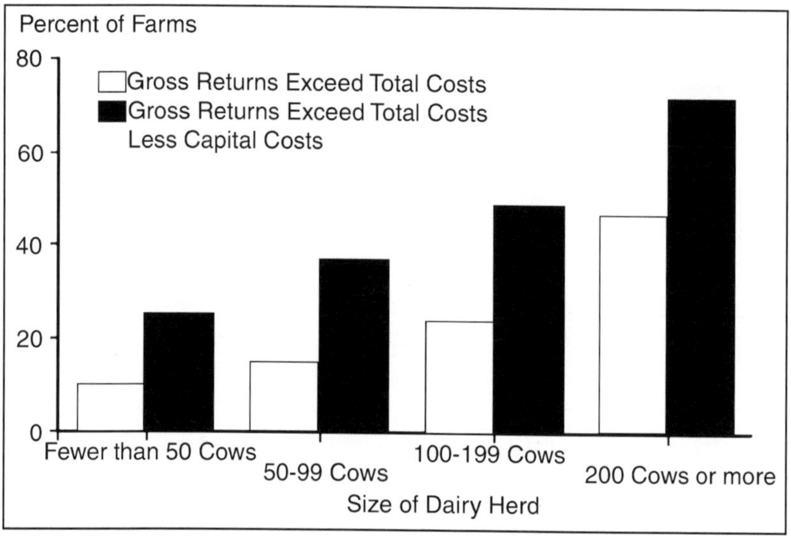

Fig. Organic Dairy Operations with Positive Returns, by Size of Operation

Notes:

Ten Per cent of the smallest and 47 per cent of the largest organic dairies have positive returns above total costs, but many more have positive returns when capital costs are excluded.

In contrast, larger organic dairies were much more likely to generate returns above capital and labour costs. These conditions suggest that organic milk production may migrate towards larger operations, as has conventional production. This production change will likely occur over an extended period as existing smaller operations use up their production facilities and equipment, ultimately facing decisions about replacing capital assets or exiting the industry.

Also, in the short term, some small operators may continue producing organic milk as a lifestyle choice despite returns that fail to cover the opportunity cost of their time.

LABOUR USE BY SIZE

Small organic dairies are at a cost disadvantage relative to larger dairies due to the opportunity cost assigned to the unpaid labour provided by operators, partners, and family members. Unpaid labour costs for organic dairies with fewer than 50 cows were $13.42 per cwt, compared with only $1.15 per cwt on dairies with 200 cows or more. If this opportunity cost is ignored, the net returns to organic milk production were more similar among the size groups. The largest dairies used substantially more hired labour than the smallest (11,530 hours versus 557 hours), but the hired labour hours per unit of production differed little among the size groups because of much higher production on the largest dairies.

The difference in unpaid labour use among the farm size groups was much less. Organic dairy operations with 200 cows or more used about 700 more hours of unpaid labour, about 20 per cent more than those with fewer than 50 cows.

The largest dairies, however, produced about 1,700 per cent more milk as a result of having more cows and an average productivity per cow that was 36 per cent higher than the smallest dairies. Enhanced productivity for these larger dairies means that unpaid labour hours per unit of production fell dramatically across the size groups from 0.82 hour per cwt on the smallest organic dairies to only 0.05 hour on the largest.

Table. Labour Use, by Size of Organic Dairy Operation, 2005

	Size of Organic Dairy			
Item	**Fewer than 50 Cows**	**50-99 Cows**	**100-199 Cows**	**200 Cows or More**
Per farm				
Hired labour hours	557	1,648	3,051	11,530
Unpaid labour hours	3,612	3,759	4,265	4,322
Total labour hours	4,169	5,407	7,316	15,852
Milk production (cwt)	4,367	8,748	15,825	78,977
Per cent				
Hired labour hours	0.13	0.19	0.19	0.15
Unpaid labour hours	0.82	0.43	0.27	0.05
Total labour hours	0.95	0.62	0.46	0.20

Unpaid labour is an opportunity cost defined as the time spent working on a dairy operation charged at a wage rate that represents what unpaid workers could earn in off-farm employment. Opportunity costs of labour may vary significantly among producers, and some producers may be willing to accept returns lower than they could earn in non-farm employment because of lifestyle preferences and costs of switching occupations, among other reasons. The incentive to earn a competitive return for the time spent on the dairy operation, however, may be a motivating force for increasingly larger organic dairies.

REGIONAL DIFFERENCES IN ORGANIC MILK PRODUCTION

Conventional milk production has historically been concentrated in the Upper Midwest and Northeast, mostly in Wisconsin, Minnesota, New York, and Pennsylvania. In the past three decades, however, milk production has expanded rapidly on large operations located in non-traditional areas of the South and West. Most prominently, California surpassed Wisconsin during the early 1990s as the number one milk-producing State.

The continued growth of dairies in these non-traditional areas is the result of a favourable climate, an abundance of alternative crops that can be used for dairy feed, their relative geographic isolation, and population growth. Although more recent in nature, regional development of organic milk production has followed a path similar to that of conventional milk production. Small organic dairy farmers and milk cooperatives in the Northeast and Upper Midwest, along with a major organic milk processor with its own large dairy farm in the West, pioneered organic milk production during the 1990s.

Organic milk production was a good fit for the small, pasture-oriented operations in the Northeast, in Wisconsin, and in other parts of the Midwest. Expansion in the West, where California and Oregon have become leading organic milkproducing States, was motivated by many of the same factors that spurred expansion of conventional milk production in this area. Figure illustrates the dramatic growth of organic milk production in the West, where the number of organic milk cows more than tripled between 2000 and 2005.

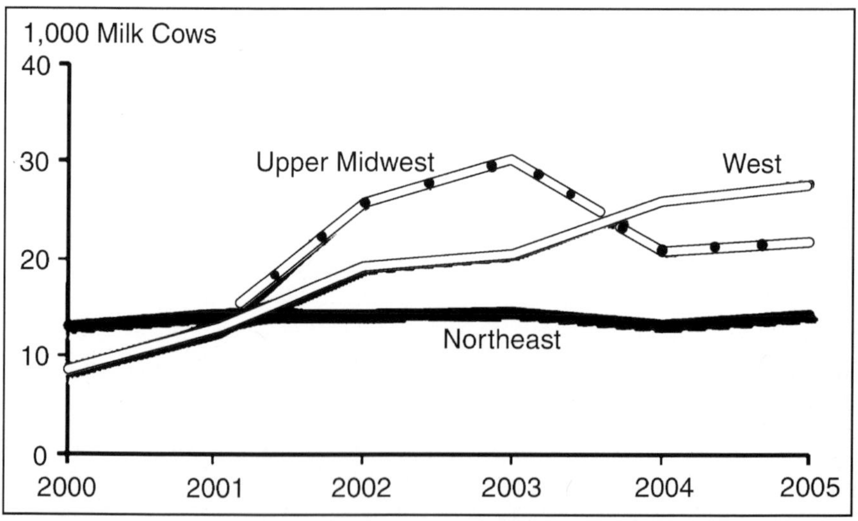

Fig. Number of Organic Milk Cows, by Region

Notes:

Between 2000 and 2005, the Number of Organic Milk Cows are more than Tripled in the West.

CHARACTERISTICS AND PRACTICES BY REGION

More than 40 per cent of organic dairies were located in both the Northeast and Upper Midwest, but these operations were small and less productive than those in the West. The Northeast includes organic dairies surveyed in New York, Pennsylvania, Maine, and Vermont; the Upper Midwest includes dairies in Wisconsin, Minnesota, and Michigan; and the West includes those in California, Oregon, Washington, and Idaho. Despite their large number, Northeast and Upper Midwest dairies accounted for a disproportionately small share of organic milk cows, 29 and 33 per cent, respectively.

Operations in the Northeast averaged 53 cows; the Upper Midwest, 64 cows; and the West, 381 cows. Organic dairy cows in the West averaged nearly 16,000 pounds of annual milk production per cow, 2,700 pounds more than in the Upper Midwest and 4,000 pounds more than in the Northeast. Organic dairies in the Northeast and Upper Midwest were more likely than dairies in the West to produce crops used for dairy cattle feed.

Nearly all farms in these regions produced hay for cattle feed, and nearly 80 per cent of farms in the Upper Midwest produced corn. Only 16 per cent of operations in the West grew corn, and just over half produced hay. As a result, the large organic dairies in the West purchased more dairy feed items than those in other regions. The percentage of farms using pasture-based feeding for dairy cattle varied little among the regions.

Table. Characteristics and Production Practices of Organic Dairies, by Region, 2005

Item	Northeast	Upper Midwest	West
Per cent of farms	44	42	7
Per cent of milk cows	29	33	31
Per cent of milk production	25	32	37
Milk cows *(number per farm)*	53	64	381
Milk production *(pounds per cow)*	11,831	13,195	15,902
Per cent of farms			
Feed crops produced:			
Corn (grain and silage)	27	79	16
Hay (alfalfa and other)	96	93	55
Operator education:			
Less than high school	37	14	5
Completed high school	43	69	48
Some college	20	17	47
Production practices:			
DHI programme participation[1]	40	60	66
Pasture-based feeding[2]	71	57	65
Artificial insemination	67	77	82
Embryo transplants or sexed semen	3	2	2
Controlled breeding/calving season	31	39	19
Regular veterinary services	36	38	46
Nutritionist services	42	44	54
Computerized milking system	2	2	6
Computerized feeding system	3	1	27

Milking parlor	28	39	100
Kept individual cow records	59	65	63
Johne's disease programme participation	14	32	35
Onfarm computer records	18	24	34
Dairy information from Internet	30	48	72
Forward-purchased inputs	9	7	21
Negotiate input price discounts	30	13	31

Notes:

[1]Dairy Herd Improvement.

[2]Pasture-based feeding is defined as providing at least half of the forage fed to milk cows during the grazing months from pasture. Organic dairies reported an average grazing period for milk cows of 6.5 months in 2005.

Operators of organic dairies in the West were, on average, more educated than those in other regions. Nearly half of farm operators in the West attended college, compared with 20 per cent or less in the other regions. More education may explain why the use of some production practices was more common on dairies in the West.

Operators of organic dairies in the West were more likely to participate in the Dairy Herd Improvement programme, use artificial insemination, use regular veterinary and nutritionist services, use computerized milking and feeding systems, access dairy information from the Internet, and forward-purchase inputs.

Higher education may contribute to a greater awareness of these practices and ease the adoption process. Also, the larger dairies in the West were able to spread the fixed costs of the investments required by some of these practices over more units of production.

PRODUCTION COSTS AND NET RETURNS BY REGION

Average operating costs were lowest for dairies in the Upper Midwest by more than $1 per cwt than in the other regions. Lower operating costs in the Upper Midwest resulted from feed costs that were about $1 per cwt less than in the Northeast and hired labour costs that were nearly $1.70 per cwt less than in the West.

Once capital costs are added, the cost difference between the Upper Midwest and West were minimal due to the impact of size economies for the larger operations in the West.

Size economies were the predominant factor in the difference between total economic costs in each region, ranging from $37.59 per cwt on the small Northeast operations to $25.13 per cwt on the large operations in the West. Most of the difference in total economic costs in the West came from unpaid labour charges that were more than $9 per cwt less than in the Northeast and about $5 per cwt less than in the Upper Midwest.

Table. Production Costs and Returns of Organic Dairies, by Region, 2005

Item	Northeast	Upper Midwest	West
Dollars per cwt sold			
Feed:	13.48	12.52	12.80
Purchased feed	6.66	3.36	9.69
Homegrown harvested feed	5.92	8.62	2.47
Grazed feed	0.90	0.55	0.64
Hired labour	1.60	1.61	3.28
Capital costs	6.31	5.44	3.67
Opportunity cost of unpaid labour	10.75	6.76	1.64
Cost summary:			
Operating costs	19.03	17.92	19.46
Operating and capital costs	25.34	23.36	23.13
Total economic costs	37.59	31.47	25.13
Gross value of production	26.83	23.56	23.47
Returns above total economic costs	-10.76	-7.91	-1.66

Notes:

Costs are defined in Appendix A: Measuring Milk Production Costs. The gross value of production includes milk sales, cull and breeding animal sales, revenue from leasing dairy animals or space to other operations, co-op patronage dividends, dairy assessment rebates or refunds, and the value of manure. The value of milk sales was determined from prices reported by the survey respondents.

A breakdown of annual feed costs per cow in each region reveals costs that were significantly less in the Northeast ($400 per cow less) and Upper Midwest ($500 per cow less) than in the West. More than half of total feed costs on organic dairies in the Northeast stemmed from homegrown feed and grazed forages, while these sources accounted for more than 70 per cent of feed costs in the Upper Midwest.

Organic dairies in these regions may have incurred lower feed costs than those shown here if their production costs were lower than the market prices charged for these inputs. In contrast, more than three-fourths of the total feed cost on organic dairies in the West was from purchased feed items. Despite higher feed costs per cow on organic dairies in the West, feed costs per cwt were competitive with organic dairies in other regions.

Total labour and capital usage were also much higher on dairies in the West, but total costs per cwt were significantly less. The cost advantage of organic dairies in the West resulted from economies of size on larger dairies and higher productivity.

Almost 50 per cent of organic dairies in the West had an average milk production per cow of 15,000 pounds or more, and only 12 per cent of organic dairies were producing less than 10,000 pounds.

In the Northeast and Upper Midwest, only a quarter of producers were producing 15,000 pounds or more, and 25 per cent of Upper Midwest producers and more than a third of producers in the Northeast had average production less than 10,000 pounds. This difference in milk yield may be due to productivity-enhancing technologies and practices more common on organic dairies in the West. Due to this cost advantage, net returns for dairies in the West were significantly greater than in the other regions.

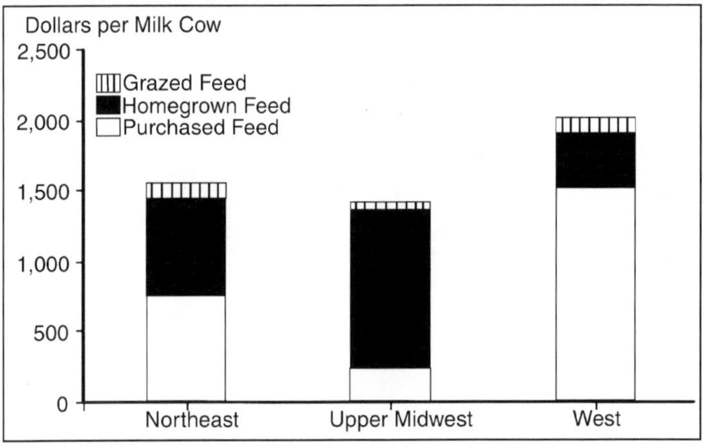

Fig. Organic Dairy Feed Costs per Cow, by Region

Notes:

Total feed costs per cow were highest for organic dairies in the West, where a much higher proportion of dairy feed was purchased than in the other regions

LABOUR USE BY REGION

Charges for unpaid labour accounted for most of the differences in total costs for organic dairies in each region. Unpaid labour costs were nearly $11 per cwt in the Northeast, nearly $7 per cwt in the Upper Midwest, but only about $2 per cwt in the West. Without these charges, the difference in total economic costs among the regions would be less than $3.50 per cwt. Total labour use for organic milk production was significantly higher in the West due to the larger operations, but labour use per cwt was much less.

Table. Labour use of Organic Dairies, by Region, 2005

Item	Northeast	Upper Midwest	West
Per cow			
Hired labour hours	24	17	26
Unpaid labour hours	76	53	12

Total labour hours	100	70	38
Per cwt			
Hired labour hours	0.20	0.13	0.16
Unpaid labour hours	0.64	0.40	0.07
Total labour hours	0.84	0.53	0.23

Total labour use per cwt in the West (0.23 hour) was less than half that in the Upper Midwest (0.53 hour) and less than a third of that in the Northeast (0.84 hour). Part of this difference in labour use per cwt was due to higher productivity in the West, but labour use per cow was also significantly less. Organic dairies in the West used 38 hours of labour per cow, compared with 100 hours in the Northeast and 70 hours in the Upper Midwest. Less labour use per cow was due, in part, to the labour-saving practices more often used on dairies in the West and to economies of size.

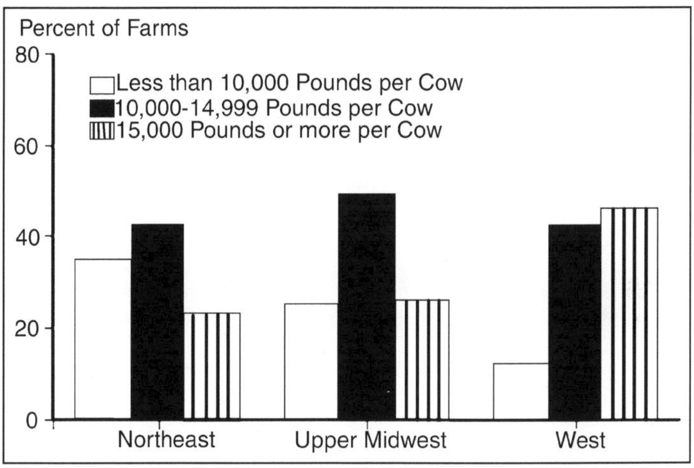

Fig. Milk Production, by Region

Notes:

Nearly half of organic dairies in the West had milk cows that produced more than 15,000 pounds per year, compared with less than 30 per cent in the other regions

PASTURE USE IN ORGANIC MILK PRODUCTION

The most common technology used on conventional dairy operations confines milk cows in large barns and limits access to pasture. This means that forage from grazing comprises very little of the forage fed to dairy cows on most conventional dairies.

Access to pasture is a requirement for the organic certification of dairy operations. Forage from grazing is also an important element of the feeding programme on many organic dairies.

Milk producers may find it easier and less expensive to maintain organic pastures for grazing dairy cows than to either purchase or produce and harvest organic crops and forage for dairy feed. Also, grazing systems may be less stressful for dairy cattle and contribute to lower veterinary expenses. Pasture-based production may also have environmental benefits, such as improved soil quality and reduced soil erosion. Pasture use for dairy feed on organic dairies, relative to conventional dairies.

More than 60 per cent of organic milk producers reported that at least half of their total forage ration came from pasture during the grazing months (an average of 6.5 months per year), compared with 18 per cent of conventional dairies. Nearly 90 per cent of organic dairies sourced at least 25 per cent of their total forage ration from pasture. In contrast, 70 per cent of conventional dairies reported that less than 25 per cent of their total forage ration fed to dairy cows came from pasture, while 40 per cent obtained none from pasture.

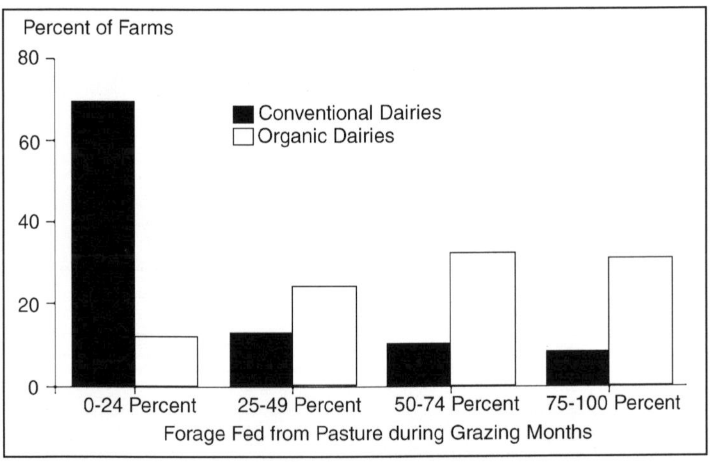

Fig. Producer Type, by Share of Dairies Forage Fed from Pasture

Notes:

More than 60 per cent of organic dairies obtained at least 50 per cent of the forage fed to dairy cows from pasture, while nearly 70 per cent of conventional dairies obtained less than 25 per cent

To evaluate the role of pasture grazing in organic milk production, surveyed producers were divided into groups that indicate the level of pasture grazing as a source of dairy cow forage during the grazing months:

- 0-24 per cent;
- 25-49 per cent;
- 50-74 per cent; and
- 75-100 per cent.

CHARACTERISTICS AND PRACTICES BY LEVEL OF PASTURE USE

Almost two-thirds of organic milk producers reported that at least 50 per cent of the forage fed to dairy cattle during the grazing months came from pasture, and a third reported that 75 per cent or more came from pasture. Only 12 per cent of organic milk producers reported that less than 25 per cent of forage fed came from pasture. Differences between organic dairies where pasture use was highest (75-100 per cent of forage fed) and lowest (0-24 per cent of forage fed) were significant and are emphasized in this section.

Farm characteristics among organic dairies in the middle groups (25-49 per cent and 50-74 per cent of forage fed from pasture) showed little difference. Operations that relied on pasture use the most were the smallest in size and produced the least milk per cow. Organic dairies with 75-100 per cent of forage fed from pasture had an average of 64 milk cows that averaged 11,289 pounds of milk per cow.

Table. Characteristics and Production Practices, by Level of Pasture use on Organic Dairy Operations, 2005

Item	Forage Fed from Pasture During Grazing Months			
	0-24 Per cent	**25-49 Per cent**	**50-74 Per cent**	**75-100 Per cent**
Per cent of farms	12	24	32	31
Per cent of milk cows	21	23	31	25
Per cent of milk production	25	23	31	21
Milk cows *(number per farm)*	135	79	78	64
Milk production *(pounds per cow)*	16,560	13,594	13,319	11,289
Region:			*Per cent of farms*	
Northeast (ME, NY, PA, VT)	21	42	43	55
Upper Midwest (MI, MN, WI)	58	44	40	35
Corn Belt (IL, IN, IA, MO, OH)	13	8	5	9
West (CA, ID, OR, WA)	8	6	12	2
Region:			*Per cent of cows*	
Northeast (ME, NY, PA, VT)	9	29	28	45
Upper Midwest (MI, MN, WI)	30	46	28	29
Corn Belt (IL, IN, IA, MO, OH)	10	9	4	8
West (CA, ID, OR, WA)	51	16	40	18
In dairy business:			*Per cent of farms*	
Less than 10 years	11	18	19	27
10 years or more	89	82	81 73	
Pasture rotation:				
At least once a day	39	48	58	72
Every 2-7 days	30	33	38	26
Less than weekly	4	1	id	1
Never	24	18	3	1
Pasture not used for feed	4	0	0	0

Notes:

id=Insufficient data for disclosure.

In contrast, dairies with 0-24 per cent of forage from pasture averaged more than twice the number of milk cows (135) and more than 5,000 pounds more milk production per cow (16,560 pounds). Thus, operations that relied less on pasture use (12 per cent) had a disproportionately large share of milk cows (21 per cent) and milk production (25 per cent), while a third of operations that relied most on pasture use accounted for only 25 per cent of milk cows and 21 per cent of production.

Organic dairies with the highest pasture use for feed were located in the Northeast. More than half of these farms were in the Northeast and operated with 45 per cent of the organic milk cows in this group. Operations in the Upper Midwest accounted for the majority of farms using the least pasture (0-24 per cent) and 35 per cent of farms using the most (75-100 per cent). Organic dairies in the West accounted for only 8 per cent of the farms using the least pasture, but these large operations held 51 per cent of the milk cows in this group.

Large operations relied less on pasture as a forage source because of the significant land requirements necessary to supply pasture for large herds. However, some large operations in the West used pasture as an important dairy feed source. Among the farms with the highest pasture use, the West (at 2 per cent) included 18 per cent of the milk cows. Organic dairies that relied the most on pasture use were in business relatively less time than other organic dairies.

Twenty-seven per cent of dairies that relied on pasture for 75-100 per cent of their forage feed had been in business less than 10 years, compared with only 11 per cent that relied on pasture for 0-24 per cent of their forage feed.

Also, farm operators who relied the most on pasture forage were younger than other farm operators.

Sixteen per cent of these farm operators were younger than 35 and only 4 per cent were older than 65, compared with 6 and 10 per cent, respectively, of the operators who used the least pasture. Younger farm operators on relatively new dairies were more likely to use pasture-based feeding for organic milk production.

Pasture as a source of forage on organic dairies was closely related to the intensity of pasture management.

Rotating pastures is a management strategy that can increase the total volume and quality of dry matter produced on pastures by allowing cows to graze for specified periods before moving them to new pastures.

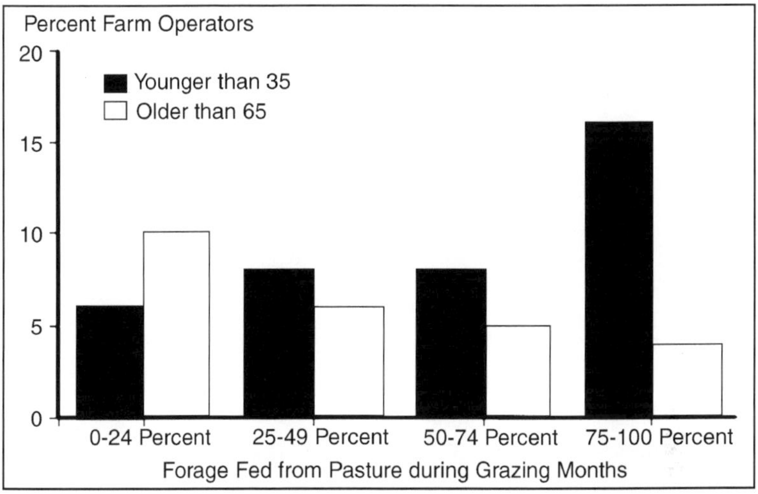

Fig. Farm Operator Age, by the Share of Dairy Forage Fed from Pasture

Notes:

Among organic farm operators that used the most pasture for dairy feed, 16 per cent were younger than 35 and only 4 per cent were older than 65

Seventy-two per cent of organic dairies that reported the highest use of pasture forage rotated pastures at least once a day, compared with less than 40 per cent of dairies that used pasture forage the least. Among dairies that used pasture the least, 24 per cent reported that pastures were never rotated and 4 per cent indicated that pasture was not a source of dairy feed.

PRODUCTION COSTS AND NET RETURNS BY LEVEL OF PASTURE USE

Substituting pasture for more expensive feed sources may appear to be an efficient way to lower dairy feed costs. Average feed costs per cwt of milk, however, were lowest on the organic operations using the least pasture for dairy forage (0-24 per cent), between $1 and $2 per cwt less than on the operations using more pasture.

Table. Production Costs and Net Returns on Organic Dairy Operations, by Level of Pasture use, 2005

Forage Fed from Pasture During Grazing Months				
Item	**0-24 Per cent**	**25-49 Per cent**	**50-74 Per cent**	**75-100 Per cent**
	Dollars per cwt sold			
Feed:	11.61	13.14	13.42	12.90
Purchased feed	7.43	5.15	7.19	6.09
Homegrown harvested feed	3.89	7.51	5.41	5.69

Grazed feed	0.29	0.48	0.82	1.13
Hired labour	2.44	1.85	2.17	2.27
Capital costs	4.11	5.36	4.48	6.70
Opportunity cost of unpaid labour	2.82	6.47	6.24	8.36
Cost summary				
Operating costs	17.74	18.55	18.99	19.44
Operating and capital costs	21.85	23.91	23.47	26.14
Total economic costs	25.33	31.53	30.64	35.99
Gross value of production	24.10	24.49	23.59	25.63
Returns above total economic costs	-1.22	-7.04	-7.05	-10.36

Notes:

Measuring Milk Production Costs. The gross value of production includes milk sales, cull and breeding animal sales, revenue from leasing dairy animals or space to other operations, co-op patronage dividends, dairy assessment rebates or refunds, and the value of manure. The value of milk sales was determined from prices reported by the survey respondents.

This relationship changes little even if charges for pasture resources are excluded from the feed costs. Average feed costs per cow declined as pasture use for dairy forage increased. Total feed costs were $1,902 per cow on organic dairies that relied on pasture for 0-24 per cent of forage fed, compared with only $1,409 on organic dairies that relied on pasture for 75-100 per cent of forage fed, a savings of about 25 per cent.

As shown previously, however, average production per cow was more than 30 per cent lower on the organic dairies that used the most pasture, thus total costs per cwt were higher.

Breaking down the distribution of milk production per cow for each group provides an indication of how milk yields declined as pasture was substituted for higher energy feed in the dairy ration. Nearly half of organic dairies feeding 0-24 per cent of forage from pasture had annual milk yields at 15,000 pounds per cow or more, compared with just 10 per cent of dairies feeding 75-100 per cent.

In contrast, more than 40 per cent of dairies using the most pasture had an annual milk yield of less than 10,000 pounds per cow, compared with 19 per cent of those using the least pasture. This suggests that improving pasture quality to achieve higher milk production can contribute to the success of organic dairies using pasture-based feeding.

The relationship between pasture use and total production costs follows much the same pattern as seen for size and production costs on organic dairies. That is, the operations using less pasture were much larger and had lower average costs per cwt than the smaller operations using more pasture.

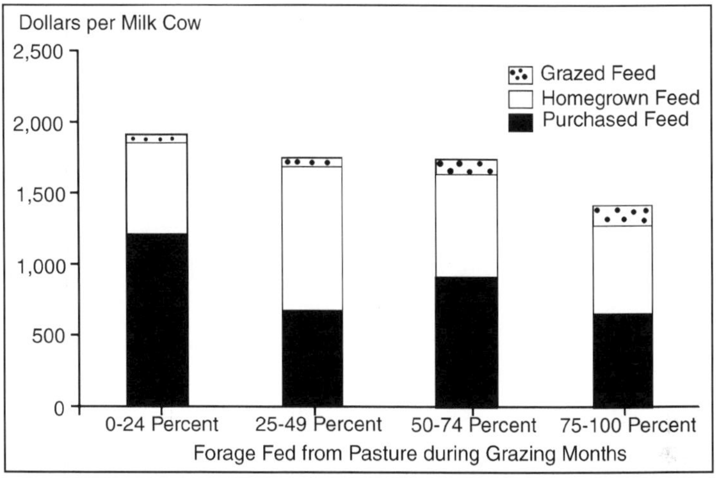

Fig. Feed Costs, by Share of Dairy Forage Fed from Pasture

Notes:

Average feed cost per cow on organic dairies were about 25 per cent less on operations using the most pasture.

Capital costs were more than $2 per cwt higher and the charge for unpaid labour was more than $5 per cwt higher on the operations where pasture forage accounted for 75-100 per cent of forage fed than on those where pasture forage accounted for 0-24 per cent of forage fed.

Larger operations were able to spread the fixed costs for capital and labour over production from more cows, and each cow was more productive on the larger operations using the least pasture for dairy forage.

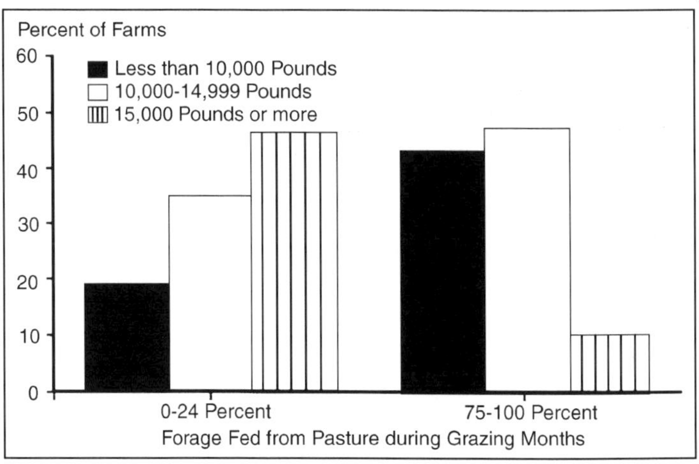

Fig. Milk Production, by Share of Dairy Forage Fed from Pasture

Notes:

Nearly half of organic dairies feeding the least pasture forage had milk yields above 15,000 pounds per cow, compared with only 10 per cent of those feeding the most pasture.

Average returns net of total economic costs were negative for dairies in each of the pasture use groups, but were significantly higher among the dairies using the least pasture. This disparity suggests that organic operations using conventional dairy feeding methods, such as confining cows and feeding higher energy feed, were more likely to generate higher returns to capital and labour resources than those relying more on pasture-based feeding.

Thus, the technologies and production methods used on organic dairies could become more like those used on conventional dairies. Changing technologies and production methods may depend on the extent to which pasture requirements for organic dairies must include pasture as part of the dairy feeding programme.

LABOUR USE BY LEVEL OF PASTURE USE

The biggest difference in organic dairy costs between those using the most and the least pasture for dairy forage was the unpaid labour charge. The labour charge on organic dairies feeding 75-100 per cent of forage from pasture was more than $8 per cwt, compared with less than $3 per cwt on dairies feeding 0-24 per cent of forage from pasture. The labour charge contributed about half the difference in total costs that were more than $10 per cwt higher for dairies using the most pasture. Table includes labour use per cow and per cwt for each level of pasture feeding.

Unpaid labour use per cwt on organic dairies feeding 75-100 per cent of forage from pasture was about three times that for those feeding 0-24 per cent of forage from pasture. Part of this difference came from fixed amounts of labour being spread over more output on the larger operations that used the least amount of pasture. Some of the labour difference, however, may be due to the pasture feeding system's being more labour intensive than handling dairy cows in confinement.

Among farms with fewer than 100 cows, the unpaid labour use on dairies feeding the least forage from pasture was still about a third less than on those feeding the most forage from pasture (0.45 hour versus 0.65 hour). In addition, total labour use among these small dairies increased steadily as more pasture was used for dairy feed.

This increase suggests that pasture-based feeding systems are inherently more labour intensive than the conventional practice of handling dairy cows in confinement, possibly due to the labour required for moving cows to, from, and between pastures.

Table. Organic Dairy Labour Use, by Level of Pasture Use, 2005

	Forage Fed from Pasture During Grazing Months			
Item	**0-24 Per cent**	**25-49 Per cent**	**50-74 Per cent**	**75-100 Per cent**
Per cow				
Hired labour hours	30	20	22	17
Unpaid labour hours	27	52	48	56
Total labour hours	57	72	70	73
Per cwt				
Hired labour hours	0.18	0.15	0.17	0.15
Unpaid labour hours	0.16	0.39	0.36	0.49
Total labour hours	0.34	0.53	0.52	0.65
Per cwt—dairies with fewer than 100 cows				
Hired labour hours	0.15	0.18	0.17	0.16
Unpaid labour hours	0.45	0.52	0.59	0.65
Total labour hours	0.60	0.70	0.76	0.81

Note:

Totals may not equal the sum of items due to rounding

COMPARING ORGANIC AND CONVENTIONAL DAIRIES

Organic dairy operations usually begin as conventional dairies that undergo the transition to become certified organic operations. Differences in farm and operator characteristics and production practices of organic and conventional dairies were identified and used to measure differences in milk production costs. These cost differences, along with estimates of organic transition costs, indicate milk price premiums that organic dairies need to be competitive with conventional milk production.

Research models were specified to describe farms that use organic production and the difference in costs between organic and conventional farms. The models accounted for the myriad of factors that influence milk production costs, such as size of operation, region, and production practices, to isolate the cost difference attributed to an organic operation. The models also accounted for the fact that organic producers are not randomly assigned among the dairy population, but instead self-select.

Self-selection could bias the cost comparison if unmeasured factors, such as the level or type of management, were correlated with both organic participation and milk production costs. This research was also designed to examine whether organic systems are more or less competitive in different segments of the U.S. dairy sector.

The models were used to evaluate the competitiveness of organic and conventional milk production for farms located in the Northeast and Upper

Midwest, those using pasture-based feeding, and among small dairies, segments of the U.S. dairy sector where organic production is most common. The results provide an indication of when organic milk production was most economical compared with conventional production.

CHARACTERISTICS AND PRACTICES OF ORGANIC AND CONVENTIONAL DAIRIES

Organic dairies were smaller than conventional dairies, averaging 82 cows per organic farm compared with 156 cows per conventional farm. The average milk production per cow was also lower on organic operations, nearly 30 per cent less. Organic operations averaged about 13,600 pounds of milk per cow, compared with nearly 19,000 pounds per cow on conventional operations. More than 80 per cent of organic dairy operations were located in the Northeast or Upper Midwest, compared with 65 per cent of conventional operations.

These regions also included 62 per cent of all organic milk cows, compared with 42 per cent of conventional milk cows. In contrast, only 7 per cent of organic dairies were located in the West, but these operations held about a third of total U.S. organic milk cows, the same as conventional operations. Organic dairies in the West (381 cows) were much larger than those in other regions and similar in average size to conventional operations (431 cows).

The average size of organic dairies in the Northeast was half that of conventional dairies (52 cows versus 104 cows), and those in the Upper Midwest were about two-thirds the size of conven tional dairies (64 cows versus 98 cows). The ARMS includes no data for organic dairy operations in the Southeast or Southwest.

Table. Characteristics of Organic and Conventional Dairy Operations, 2005

Type of Dairy		
Item	**Conventional**	**Organic**
Milk cows (number per farm)	156	82
Milk production (pounds per cow)	18,983	13,601
Per cent of farms/cows		
Region:		
Northeast (ME, NY, PA, VT)	26/17	44/29
Upper Midwest (MI, MN, WI)	39/25	42/33
Corn Belt (IL, IN, IA, MO, OH)	15/10	7/7
Southeast (FL, GA, KY, TN, VA)[1]	6/6	0/0
Southwest (AZ, NM, TX)[1]	2/10	0/0
West (CA, ID, OR, WA)	11/32	7/31
Milk cows per farm		
Region:		
Northeast (ME, NY, PA, VT)	104	52

Upper Midwest (MI, MN, WI)	98	64
Corn Belt (IL, IN, IA, MO, OH)	108	75
Southeast (FL, GA, KY, TN, VA)[1]	152	0
Southwest (AZ, NM, TX)[1]	781	0
West (CA, ID, OR, WA)	431	381
Per cent of farms		
Farm operator:		
Off-farm occupation	2	4
Education:		
Less than high school	18	26
Completed high school/some college	66	54
Graduated from college	16	20
Age (years)	51	49
In dairy business (years)	23	21
Selling certified organic milk (years)	na	5
Exit dairy business:		
5 years or less	25	16
10 years or less	51	33
20 or more years	30	47

Notes:

na=Not applicable. [1]Organic dairies were not surveyed in the Southeast and Southwest regions.

Most farm operator characteristics were similar among conventional and organic dairies. Nearly all farm operators in both groups reported farming as their primary occupation, common for dairy operations due to their substantial onfarm labour requirements. The distribution of operator education suggests that those in the organic group were neither more nor less educated than operators of conventional dairies.

The average age of farm operators on conventional and organic dairies was similar. Most organic operations converted to organic production after years of conventional production. Of the 21 years that organic operators reported being in business, organic milk was produced for only 5 years. Organic producers were more optimistic about their future, as significantly fewer planned to exit in the next 10 years and significantly more planned to be in business for 20 or more years.

The primary difference in the production practices used by organic and conventional dairies was found in their feeding systems. More then 60 per cent of organic operations reported using pasture-based feeding that provided more than 50 per cent of seasonal forage from pasture (during the grazing months), compared with just 18 per cent of other operations.

Although not available to organic producers, rBST was used by 17 per cent of conventional operations that were also more likely to utilize regular veterinary services and a nutritionist to formulate dairy rations. Differences in production practices may have contributed to the higher production per cow on conventional versus organic operations. Labour use also distinguished conventional from organic dairies.

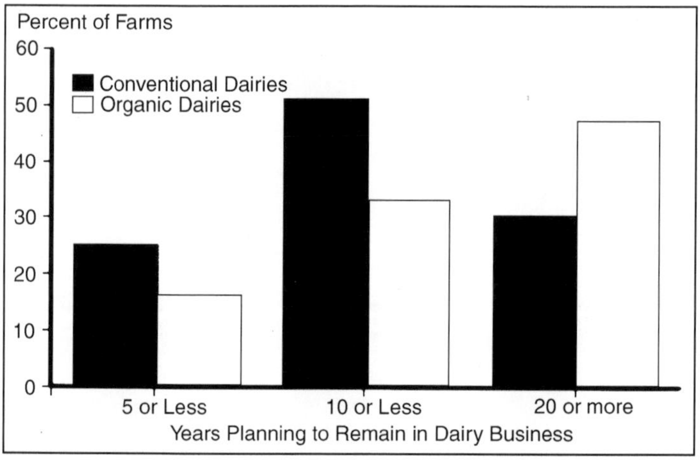

Fig. Plans to Exit the Dairy Business, by Type of Producer

Notes:

Operators of organic dairies were more optimistic about their future.

Organic operations used nearly twice the hours of total labour per cwt, on average, than did conventional operations (0.50 hour versus 0.26 hour). Most of the labour difference came from significantly more unpaid labour hours worked on organic dairies and was influenced by the smaller average size and lower productivity of organic dairies. Fixed labour inputs were spread over fewer units of production on the smaller organic dairies.

FACTORS AFFECTING THE USE OF ORGANIC MILK PRODUCTION

Several operator and farm characteristics were statistically associated with the organic approach to milk production. Among operator characteristics, education and planning horizon were important. Dairy farmers graduating from college were more likely to be organic producers.

Table. Production Practices and Labour Use on Organic and Conventional Dairy Operations, 2005

Type of Dairy		
Item	**Conventional**	**Organic**
Per cent of farms		
Production practices		
DHI programme participation[1]	45	46
Pasture-based feeding[2]	18	63
Milking three times or more daily	7	1

rBST (recombinant bovine somatotropin)	17	0
Artificial insemination	82	73
Embryo transplants or sexed semen	10	3
Controlled breeding/calving season	25	35
Regular veterinary services	69	38
Nutritionist services	72	45
Computerized milking system	5	2
Computerized feeding system	7	3
Milking parlor	50	41
Keep individual cow records	61	62
Johne's disease programme participation	20	26
Onfarm computer records	26	21
Dairy information from Internet	38	41
Forward-purchased inputs	20	8
Negotiate input price discounts	35	21
Hours per cwt		
Labour use:		
Paid labour	0.12	0.16
Unpaid labour	0.13	0.34
Total labour	0.26	0.50

Notes:

[1]Dairy Herd Improvement.

[2]Pasture-based feeding is defined as providing at least half of the forage fed to milk cows during the grazing months from pasture. Organic dairies reported an average grazing period for milk cows of 6.5 months in 2005.

Dairy operations planning to exit the industry in the next 10 years were less likely to be organic, indicating that operations with a longer planning horizon were more likely to use the organic approach. More educated dairy farmers with long-term plans to remain in business are probably more willing and able to make the necessary investments or take the additional risks associated with organic production.

Size and location of dairy operations were also important factors influencing operators to go organic. The likelihood of being organic decreased as the number of milk cows on a farm increased. Larger operations had less incentive to go organic because of economies of size in milk production, pasture requirements for organic certification, and possible difficulties sourcing large quantities of organic inputs.

The proximity and accessibility of grazing land to the operation may also have been important because barriers, such as highways and streams, may limit available pasture.

Location in the Northeast or Upper Midwest was also associated with a higher probability of being organic. The organic dairy industry began in these areas and may offer a more developed infrastructure for handling organic milk.

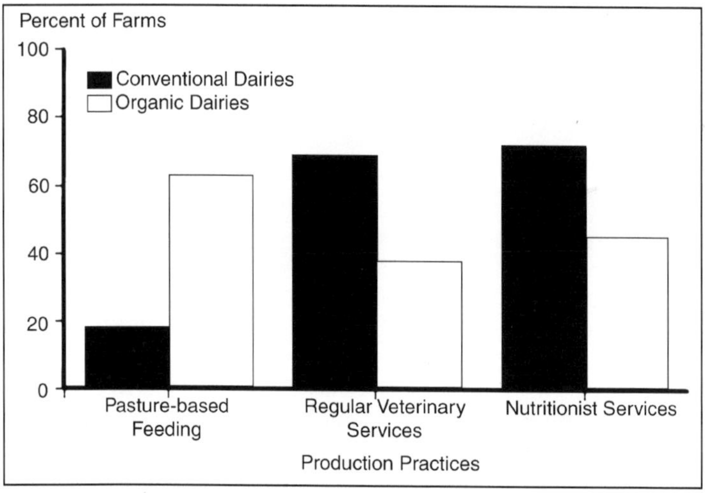

Fig. Production Practices, by Type of Producer

Notes:

Organic operations used pasture-based feeding more often, while conventional operations used regular veterinary and nutritionist services more often.

Operations with a pasture-based feeding programme were more likely to be organic, possibly because the land base enabled them to meet organic pasture requirements and pasture was more easily managed organically than field crops.

Results suggest that using pasture as a significant source of dairy feed was one of the best predictors of being an organic dairy.

Location in counties with a greater concentration of farms with milk cows included more organic production, potentially as a result of organic processors who chose to locate and recruit in areas with a large concentration of producers in order to reduce milk transportation and other transaction costs.

FACTORS AFFECTING MILK PRODUCTION COSTS

This section of the report examines the relationship between farm and operator characteristics and three levels of production costs—operating costs, operating and capital costs, and total economic costs. Older farm operators had higher costs than younger operators.

Operating, operating and capital, and total economic costs all declined as farm size increased, consistent with economies of size. Costs declined as size increased, at a decreasing rate, as fixed capital and labour costs were spread over more units of output. Significant economies of size with respect to capital and labour were expected, but operating costs also declined with size, possibly due to greater feed efficiency or lower prices paid for feed items on larger farms. Farm location influenced production costs as dairies in most other regions had lower costs than those in the Northeast. Technology use was also an important determinant of production costs. Farms using more technology had lower per unit capital and labour costs, possibly by increasing their productivity.

Pasture-based feeding had a negative, but insignificant, effect on operating costs. Feed costs were lower for organic dairies that substituted pasture for other feed items, but lower production from pasture-fed cows offset the cost savings. Pasture-based feeding was associated with higher total economic costs due to higher labour requirements.

After accounting for factors that influence production costs, operating costs for organic production were $4. 78 per cwt higher, operating and capital costs were $5.65 per cwt higher, and total economic costs were $6.79 per cwt higher than for conventional production among all U.S. dairies.

Results were similar for dairies in the Northeast and Upper Midwest and among small farms (less than 150 cows). The estimated difference in production costs between pasture-based organic and pasture-based conventional farms, however, was much less than among all dairies ($2.87 per cwt for operating costs, $3.00 per cwt for operating and capital costs, and $3.57 per cwt for total economic costs). Therefore, it is not surprising that many pasture-based dairies use organic production.

ORGANIC TRANSITION COSTS

The estimated cost differences indicate the additional costs incurred by operations producing organic milk relative to conventional operations but do not include the costs associated with the transition period. Data from the ARMS did not indicate the actual costs incurred during transition, so the estimated cost differences between organic and conventional milk production were approximated for the transition period.

Before an operation is certified to sell organic milk, pasture and cropland for dairy feed must be managed organically for a minimum of 36 months, and the dairy herd must be fed and managed organically during the last 12 months of that period.

As a result, organic operations must undergo 3 years of higher costs before the higher organic milk prices are received.

Higher costs for 3 years are a necessary investment to return higher milk prices over the expected life of the operation. This initial investment is determined by the estimated additional costs incurred by organic operations over the 3-year transition period. During year 3, when both the land and dairy herd must be managed organically, the total additional costs were charged. During years 1 and 2, when only the land is managed organically, 50 per cent of the additional costs were charged.

Table. Estimated Additional Costs Incurred by Organic Dairy Farms Compared with Conventional Dairy Farms, 2005

Economic	Operating Costs	Operating and Capital Costs	Total Costs
Dollars per cwt sold			
All farms:			
Producing organic	4.78	5.65	6.79
Transitioning to organic	na	0.72	0.86
Total additional costs	4.78	6.37	7.65
Northeast and Upper Midwest farms:			
Producing organic	4.51	5.43	6.77
Transitioning to organic	na	0.69	0.86
Total additional costs	4.51	6.12	7.63
Pasture-based farms:			
Producing organic	2.87	3.00	3.57
Transitioning to organic	na	0.38	0.45
Total additional costs	2.87	3.38	4.02
Small farms (fewer than 150 cows):			
Producing organic	4.67	5.78	7.82
Transitioning to organic	na	0.73	0.99
Total additional costs	4.67	6.51	8.81

Notes:

na=Not applicable.

Transition costs were treated as a capital investment necessary to return the higher organic milk price over the expected life of the operation and thus were not part of annual operating costs.

This corresponds with about half of the feed cost on organic dairies from homegrown supplies, which would be managed organically during the full 3 years. The annualized cost of this investment was computed by spreading it over an expected operating life of 20 years. The estimated transition costs and total additional costs on organic operations are shown in table.

Transition costs were $0.72 per cwt for operating and capital costs and $0.86 per cwt for total economic costs. Thus, the total estimated additional costs among U.S. dairy farms for producing organic relative to conventional milk were estimated at $4.78 per cwt for operating costs, $6.37 per cwt for operating and capital costs, and $7.65 per cwt for total economic costs. Among pasture-based dairies, the cost differences were much lower at an estimated $2.87 per cwt for operating costs, $3.38 per cwt for operating and capital costs, and $4.02 per cwt for total economic costs.

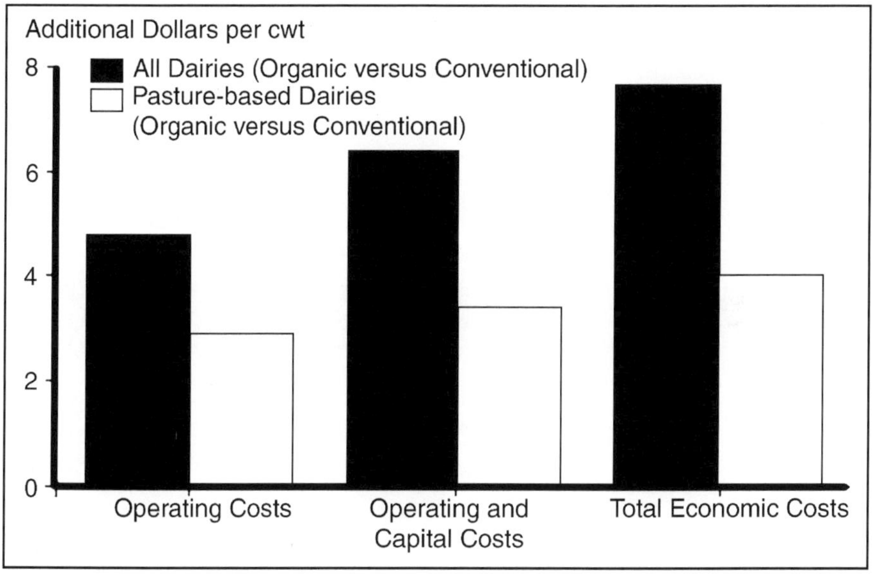

Fig. Additional Costs of Organic versus Conventional Milk Production

Notes:

The additional costs of organic compared with conventional milk production were much lower among pasture-based dairies than for all dairies.

CHALLENGES OF ORGANIC MILK PRODUCTION

Organic milk producers were asked in the ARMS what they considered the most difficult aspect of organic milk production. Forty per cent of producers reported that certification paperwork and compliance cost were the most challenging.

Sourcing organic inputs, including grains and forages, feed supplements, and replacement heifers, was the most difficult aspect reported by 23 per cent of organic producers. High costs of production and maintaining animal health were challenging aspects reported by 17 and 13 per cent of producers, respectively.

In this section, the challenges of organic milk production are explored by examining how reported challenges to organic milk production varied across the sector. Producer reports were summarized by operation size, region, level of pasture use, and operator characteristics to see what factors may constrain the adoption and growth of organic milk production.

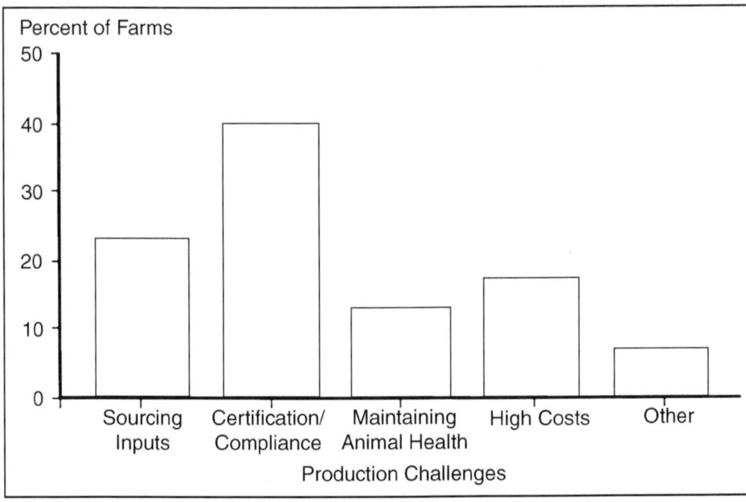

Fig. Challenges of Organic Milk Production

Notes:

40 per cent of organic dairies cited certification paperwork and compliance costs as the most difficult aspect of organic production.

FARM CHARACTERISTICS AND ORGANIC PRODUCTION CHALLENGES

The challenges reported for organic production varied little among dairy size groups, except for the largest dairies. About a third of the largest organic dairies, those with 200 cows or more and an average size of nearly 500 cows, reported sourcing inputs as the most difficult aspect, compared with 20-25 per cent of smaller dairies. This indicates that sourcing organic inputs in volume may be difficult for some large organic producers and may be a factor constraining the size of organic operations. Smaller dairies reported certification/ compliance as the most difficult aspect, and the smallest dairies, those with fewer than 50 cows, were the least likely to report animal health and production costs as most challenging.

Table. Most Difficult Aspect of Organic Milk Production, by Farm and Operator Characteristics, 2005

Most Difficult Aspect				
Maintaining				
Item Other	Sourcing Inputs	Certification/ Compliance	Animal Health	High Costs of Production
Per cent of farms reporting aspect				
Size:				

Fewer than 50 cows	25	44	10	13	8
50-99 cows	20	36	15	21	8
100-199 cows	20	41	19	20	0
200 cows or more	34	30	17	17	2
Region:					
Northeast	19	32	13	30	6
Upper Midwest	24	51	10	6	9
West	51	id	14	20	id
Pasture use:					
0-24 per cent	13	43	15	16	13
25-49 per cent	22	47	id	17	id
50-74 per cent	26	42	14	13	5
75-100 per cent	24	31	13	21	11
Operator age:					
Younger than 50 years	20	43	16	14	8
50 years or older	26	37	10	21	6
Operator education:					
Less than high school	17	54	4	18	8
Completed high school	22	40	17	15	7
Some college	33	22	16	21	8
Organic experience:[1]					
1 year	26	33	25	11	5
2-4 years	26	43	12	13	7
5 or more years	19	39	12	22	8

Notes:

id = Insufficient data for disclosure.

[1]Number of years an operator has produced organic milk.

Organic milk production challenges were different in each region. The most difficult aspect for organic dairies in the Northeast was split between certification/compliance and production costs at a third each. Dairies in the Northeast were the least likely to report sourcing inputs as most challenging, possibly because these small dairies often produce organic dairy feed or can buy it locally. Half of Upper Midwest dairies reported certification/compliance, while half of organic dairies in the West indicated sourcing inputs as the most difficult aspect of organic milk production.

The volume of organic inputs needed on large farms in the West may account for the level of concern with sourcing inputs. The most difficult aspect of organic milk production did not differ much among the pasture use groups, except for those using the most pasture.

Dairies that relied on pasture for 75-100 per cent of forage during the grazing months were less likely to report certification/compliance as the most difficult aspect of organic production. Certification/compliance may be less difficult for these dairies because the feeding programme satisfies the pasture requirement for organic certification. More than 40 per cent of the dairies in each of the other pasture use groups reported certification/compliance as the most difficult aspect of organic milk production.

OPERATOR CHARACTERISTICS AND ORGANIC PRODUCTION CHALLENGES

Responses based on operator age varied by 6-7 percentage points. Operators younger than 50 years of age were more concerned with certification/compliance and maintaining animal health. Those older than 50 years of age reported sourcing inputs and production costs as the most difficult aspects of organic milk production. Certification/compliance was reported as the most difficult aspect of organic milk production by more than half of farm operators with less than a high school education. As operator education increased, certification/compliance was less of a concern.

Forty per cent of operators with a high school education, compared with only 22 per cent of those with some college, reported certification/compliance as the most difficult aspect. Results suggest that education may ease the certification process. In contrast, sourcing inputs was reported as most challenging with more frequency as operator education increased.

The relationship between organic experience and the challenge of organic production was examined by grouping farm operators by the number of years they had produced organic milk.

Dairy farm operators in their first year of organic production were more likely to indicate that maintaining animal health was most challenging, reported by 25 per cent of operators.

Organic milk producers may have learned quickly to manage animal health as these concerns fell to 12 per cent by year 2 and beyond. High production costs were an increasing concern as organic operators became more experienced, but certification/compliance issues were still the primary concern for organic dairy operators regardless of their experience.

CONCLUSIONS

Unique and detailed data from a 2005 survey of U.S. dairy operations were used to characterize organic dairies and to compare them with conventional dairies. The dataset is unique in that it includes a targeted sample of organic producers at a much higher rate than their occurrence in the population of all dairy farms. This targeted sample allows for an examination of the structure and costs associated with organic milk production and an analysis of differences between the conventional and organic sectors. Operation size was a primary determining factor for a dairy operation being organic.

Small farms may view the organic approach as an alternative by which to reorganize current resources to improve farm returns and the odds of economic survival. Small-scale production may also be more conducive to sourcing organic inputs, which may be limited in some areas. Large dairies have more invested in production technologies that typically confine milk cows to large barns and limit access to pasture and are able to take advantage of economies of size. Thus, large farms may have less incentive to consider production alternatives.

Further, large farms may have greater difficultly sourcing sufficient quantities of organic inputs, and transitioning to organic production may require more adjustments due to pasture certification requirements. Large organic dairies more often reported sourcing inputs and organic certification/compliance requirements as the most challenging aspects of organic milk production. Dairies were more likely to produce milk organically if they were located in the Northeast or Upper Midwest. These areas have a long history of small dairies and thus a successful infrastructure to provide inputs and manage output from several small operations.

The largest U.S. organic milk cooperative pioneered organic milk production in the Northeast and Upper Midwest during the mid-1990s. Proximity to markets with highly affluent consumers also made these regions attractive to organic milk operations. Access to pasture for dairy feed also had a strong influence on whether a dairy becomes organic. Operations using pasture-based feeding satisfy the pasture requirements for organic certification, and organic pasture management is generally easier than organic crop management and less costly than purchasing organic dairy feed.

Fewer small, pasture-based organic dairies were as concerned about certification/compliance requirements as other organic dairies. Results from a statistical model comparing conventional and organic milk production costs indicated that average operating costs for organic dairies were $4.78 per cwt higher, and operating and capital costs were $5.65 per cwt higher after accounting for other factors that influence production costs.

Including an estimate of the additional costs incurred during transition, average organic milk production costs were $4.78 and $6.37 per cwt higher, respectively.

With an average price premium of $6.69 per cwt for organic milk, organic milk producers, on average, covered the additional operating and capital costs of organic production in 2005. Most organic dairies were small operations that used primarily unpaid operator and family labour. Returns above operating and capital costs on these small organic operations compared favourably with those of small conventional operations, suggesting that there may be economic incentives for some small existing dairies that have already committed much of the fixed investment to consider the transition to organic milk production.

Additional economic costs for organic production averaged nearly $1 per cwt more than the organic milk price premium in 2005. Thus, low returns to unpaid labour and management may limit startup organic dairies unless they can enter the industry at a much larger scale of production than the current norm. Even though most organic dairies were much smaller than conventional dairies, economic forces suggest that similar economies of size were available to organic producers. Some small organic dairies were earning enough to operate, but many were not able to cover the opportunity costs of investments in capital and the operator's time compared with larger operations.

These economic conditions suggest that organic milk production may migrate towards larger operations, as has conventional production. This production shift will likely occur over an extended period as many existing small organic operations either choose to replace their capital assets with larger production units or exit the industry. In the short term, some small operators may continue producing organic milk as a lifestyle choice, despite generating returns that do not cover the opportunity cost of their time.

Organic milk farms able to use pasture resources for a significant portion of dairy feed were very competitive with pasture-based conventional producers. Average operating costs for pasture-based organic milk production were estimated to be about $3 per cwt higher than those for pasture-based conventional production, and total economic costs for organic pasture-based milk production were about $4 per cwt higher than for pasture-based conventional production, significantly less than the average organic milk price premium in 2005.

Economic incentives appear to favour pasture-based dairies transitioning to organic production, and possibly startup organic dairies that can take advantage of pasture resources suitable for organic dairy feed. While the economics of pasture-based organic milk production compared favourably with that of pasture-based conventional production, pasture-based organic dairies had lower average milk production per cow and higher per unit costs than other organic dairies.

This suggests that organic operations using conventional dairy feeding methods, such as confining cows and feeding harvested forages, are more likely to generate higher returns to capital and labour than those using pasture-based feeding. Thus, the technologies and production methods used on organic dairies may become more like those used on conventional dairies.

Interpretation and implementation of revised organic pasture rules may shape the future structure of the organic industry. If pasture rules become more stringent, the challenges of certification/compliance could have a major impact on the sizes and types of farms able to produce certified organic milk.

This report attempts to shed light on the structure of organic milk production, factors affecting whether a dairy becomes organic, and the relative costs and returns of conventional and organic milk production.

Based on these findings, implications for the structure of organic milk production were developed. However, conclusions derived from the analysis were based on 2005 organic and conventional milk and feed price relationships and could change with adjustments in relative milk prices, markets for conventional and organic inputs, technological improvements in dairy farming, and new organic pasture regulations.

Bibliography

A K Kolay: *Manures and Fertilizers*, Atlantic Publication, Delhi, 2007.

A. Daniel Hall: *Fertilizers and Manures*, Biotech Books, Delhi, 2002.

A.P. Jaiswal and Deo Kant Prasad: *Handbook of Soil Fertilizer and Manure*, Enkay Publishing House, Delhi, 2012.

A.P. Ruhil, T.K. Mohanty and S.S. Lathwal: *Data Mining Techniques for Farm Animal Management*, Agrotech Publishing Academy, Delhi, 2013.

Archana Ruhela: *Agriculture and Food Security*, Oxford Book Company, Delhi, 2008.

Bruno Dorin and Frederic Landy: *Agriculture and Food in India : A Half-Century Review, From Independence to Globalization,* Manohar Publication, Delhi, 2009.

Dalip Kumar and Asmi Raza: *Agriculture and Food Security : Contemporary Issues*, Deep and Deep Publication, Delhi, 2011.

Diwakar, D.M.: *Agriculture and Industry: Dynamics of Imbalances in India*, Manak Publication, Delhi, 1992.

E.J. Russell: *Handbook on Soils and Manures*, Discovery Publishing House, Delhi, 1999.

F R Marshall: *Breeding Farm Animals*, Asiatic Publication, Delhi, 2006.

Gauri Shanker Dokania and Manisha Dokania: *Economics of Dairy Farming : A Micro Analysis*, Satyam Publishing House, Delhi, 2012.

Harmeet Singh: *Dairy Farming*, APH Publication, Delhi, 2011.

James A.S. Watson, Wattie J. Mills, K.L. Blaxter and C.C. Thiel: *Farm Animals and their Management*, Satish Serial Publication, Delhi, 2005.

John Hammond: *Farm Animals : Their Breeding Growth and Inheritance*, Axis Books, Delhi, 2013.

K.N. Raja, P. Kathiravan, Rekha Sharma and B.K. Joshi: *Genome Analysis in Farm Animals*, Jaya Publishing House, Delhi, 2013.

M S Shaktawat; Mohan Singh; N C Aery; B N Swami and M K Katewa: *Phosphate Rich Organic Manure : An Alternate to Phosphatic Fertilizers*, Himanshu Publication, Delhi, 2004.

M. Singh, D.K. Sharma and U.K. Mishra: *Organic Dairy Farming*, Satish Serial Publication, Delhi, 2011.

Mohd. Kalimuddin: *Milk Analysis and Dairy Farming*, Enkay Publication, Delhi, 2012.

Mridula Mishra: *Agriculture and Food Economics*, Serials Publication, Delhi, 2010.

P K Behera: *Plant Compost-Manure and Agro-Chemicals Analysis : A Laboratory Manual (2 Vols-Set)*, Dominant Publication, Delhi, 2007.

Pratap Mal: *Agriculture and Rural Development*, Mohit Publication, Delhi, 2001.

Rajendra Reddy and J.P. Abhay Shankar: *Agriculture and Rural Development*, Commonwealth Publication, Delhi, 2008.

Ramniwas Sharma: *Agriculture and Food Safety*, Biotech Books, Delhi, 2006.

Rana Ranjeet Singh and Md. Manzarul Islam: *Farm Animal Management: Principles and Practices*, New India Publishing Agency, Delhi, 2014.

S S Kundu; Sultan Singh; S K Mahanta and G H Pailan: *Feeding Farm Animals*, Satish Serial Publication, Delhi, 2005.

Santaniello; R E Evenson; D Zilberman and G A Carlson: *Agriculture and Intellectual Property Rights : Economic, Institutional and Implementation Issues in Biotechnology*, Universities Press, Delhi, 2003.

Shyam Kishore Singh: *Dairy Farming*, Alfa Publications, Delhi, 2012.

Uma Shankar Singh: *Dairy Farming*, Anmol Publication, Delhi, 2008.

Vir Singh and Babita Bohra: *Dairy Farming in Mountain Areas*, Daya Publication, Delhi, 2006.

Y.G. Dugwekar: *Reproductive Biotechnology of Farm Animals*, Agrotech Publication, Delhi, 2006.

Y.V.R. Reddy, M. Jayachandra Reddy, B. Hemalatha and Y.S. Ramakrishna: *Economic Evaluation of Dairy Farming in Rural India*, Agrotech Publication, Delhi, 2006.

Index

H

I

L

M

N

O

P

Q

R

S

T

V

W

Z